Identity-Based Encryption

Sanjit Chatterjee • Palash Sarkar

Identity-Based Encryption

Springer

Sanjit Chatterjee
Department of Computer Science
and Automation
Indian Institute of Science
Bangalore
India 560012
sanjit@csa.iisc.ernet.in

Palash Sarkar
Applied Statistics Unit
Indian Statistical Institute
203 B.T. Road
Kolkata
India 700108
palash@isical.ac.in

ISBN 978-1-4899-9697-8   ISBN 978-1-4419-9383-0 (eBook)
DOI 10.1007/978-1-4419-9383-0
Springer New York Dordrecht Heidelberg London

Printed on acid-free paper

Springer is part of Springer Science+Business Media (www.springer.com)

To our families

Preface

One of the main reasons for writing this monograph is the curiosity to know more. Identity-based encryption is a fascinating area of modern cryptography. We have done some work in this area and wanted to explore the area in some depth. This motivation, however, was not sufficient for us to take up the job. The catalytic effect is due to Peter Wild, who suggested us to take up this work and very kindly referred us to Springer.

Determining the focus of a monograph is difficult. More specifically, it is difficult to determine who would benefit from the text. Within a decade or so, identity-based encryption has emerged as a distinct area of research starting, virtually, from scratch. Naturally, IBE now attracts many research students from across the world. For a beginner, it might be a little difficult to get a unified account of the evolution of this research area from the scattered literature. It is precisely this gap that we try to fill through this monograph.

Starting from the basic ideas, the monograph attempts to cover all the important IBE schemes that have been proposed till date. Among these are included the more recently proposed lattice-based IBE schemes. A major thrust of the research in IBE is the so-called security proof. Proofs of the security reductions of most of the important schemes are provided in details. Our guiding principle in selecting the proofs for inclusion is that they highlight some novel techniques. Many more schemes are described without any formal proof and in some cases an intuitive description of the security reduction is provided. Hopefully, going through the book will help a research student to grasp the central ideas involved in constructing IBE schemes. If this is indeed achieved, then our efforts will have succeeded. We also hope that the book may be useful to experts as reference for quickly looking up a particular topic.

The actual writing of the monograph took several months (and included missing a deadline). We are thankful to Springer, especially Susan Lagerstrom-Fife for her enthusiasm and support to the project. Jennifer Maurer looked over the manuscript several times and provided useful feedback.

Technical feedback on the contents of the monograph were provided by several of our colleagues and students. In fact, they formed the test cases for the judging the usefulness of the monograph. We gratefully acknowledge the comments and feedback received from Rana Barua, Sanjay Bhattacherjee, Sherman S. M. Chow, M. Prem Laxman Das, Kishan C. Gupta, Koray Karabina and Somindu C. Ramanna. Without their feedbacks, there would surely been many more mistakes present in the book than there is now. Needless to say, any errors that do remain are entirely our own responsibility.

Kolkata, India,   *Sanjit Chatterjee*
November 2010   *Palash Sarkar*

Contents

Chapter 1
Introduction

1.1 Background

Science, it has been argued by Kuhn in 1962 [126], advances through paradigm shifts. Concepts emerge that open-up new vistas of research, fundamentally changing the way we are used to looking at things. Between these paradigm shifts remain the periods of consolidation. Periods when human mind explores the newly found territory, shedding light on hitherto unknown dimensions. If radical changes are the hallmarks of paradigm shifts, the period within witnesses small but continuous developments, occasionally marked by its own milestones. It is in these periods that human faculty tries to grasp the full significance of the new concepts, consolidates its gains and thereby pushes the boundary of our collective knowledge further. The prospects, nevertheless, bring with it new problems too. Perhaps, by the way, making ground for the next paradigm shift.

Cryptology, as a branch of human knowledge, is no exception to this common story. Whenever civilisation reached a certain level of sophistication, the need for secret communication between two geographically distant parties has arisen. Politics, military and business are the three dominant areas of human activity where such communication becomes essential. With the gradual evolution of technology from ancient days to modern times, several innovative forms of encryption methodology have been developed, extensively used and then discarded after brilliant insights into new cryptanalytic methods. The cycle of development and analysis of cryptographic schemes have thus progressed to become a fascinating subject of human intellectual endeavor. There are excellent accounts of this historical development in [115, 164].

The basic problem of cryptography can be considered to be the issue of secure communication between two parties using a communication medium which is not under the exclusive control of these two parties. Such a medium is called a public channel, highlighting the fact that the information flowing across this channel is publicly accessible. An alternate view of cryptography is the task of building an implicit secure communication channel over an explicitly given insecure public channel. Figure 1.1 provides an overview of the classical model of performing this task.

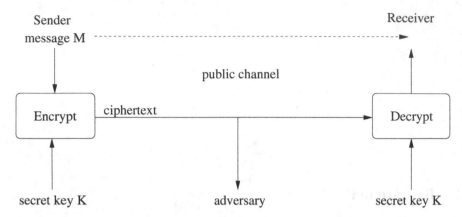

Fig. 1.1 An overview of symmetric key encryption.

Both the sender and the receiver share a common secret key K which is not known to the adversary. The ciphertext moving across the public channel is a function of the message and the secret key K. An adversary has access to the ciphertext, but, without knowledge of K should be unable to obtain the intended message M. The receiver, on the other hand, knows K and should be able to recover M from the ciphertext.

Since antiquity, human mind had accepted this model as the natural and perhaps the only model of secure communication. There had been no reason or motivation to look beyond this model. Things, however, changed with the advent of radio communication. This enabled large-scale communication. But, the full functionality of radio communication could not be realised if the confidentiality of the communication could not be assured. Thus, arose the problem of ensuring secure communication between any two of a number of parties.

Suppose there are 100 parties. Using the classical model, secure communication between any two parties requires a secret key per pair of parties. So, the total number of secret keys in the system is $\binom{100}{2}$ and each party has to maintain 99 secret keys. See Figure 1.2 for a pictorial representation of this scenario. To visualise the immensity of the problem, one may change 100 to any number n that would be practical in the real world.

Necessity is the mother of invention (attributed to the Greek philosopher Plato) and this is exemplified in the further development of cryptography. While previously, there had been no reason to consider anything but the classical model, with the advent of radio and the concomitant revolution in communication, arose the necessity of developing manageable methods of ensuring security of such communications. Solution to this problem was made possible through a paradigm shift in the discipline of cryptology.

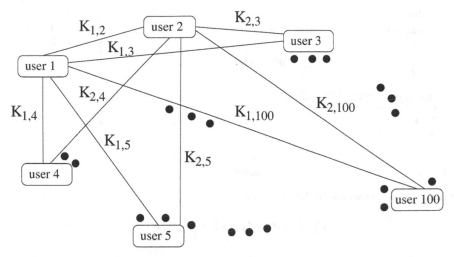

Fig. 1.2 Secure communication among 100 users using the classical model.

The basic idea of the new paradigm was simple. Instead of using the same key for encryption and decryption, one may consider two separate keys for each party – one key is used for encryption and the other key is used for decryption. The encryption key may be made public, so that any other party (say Alice) may send an encrypted message. The decryption key, on the other hand, should be kept secret, so that only the intended receiver (say Bob) can decrypt the ciphertext. See Figure 1.3 for an overview of this idea. This is called public key encryption (PKE).

Though it is surprisingly simple, the idea eluded researchers for a long time. This perhaps lends credence to Kuhn's theory of major scientific developments (alluded to in the opening paragraph) as proceeding through paradigm shifts. The human mind had grown accustomed to the belief (or paradigm) of using the *same* key for both encryption and decryption and hence, found it extremely difficult to conceive the shift where the encryption and decryption keys are different.

It was first published by Diffie and Hellman in their seminal paper [77] appropriately titled "New Directions in Cryptography". Somewhat interestingly, researchers working for the British government had also obtained the same idea but, their work remain classified for several decades. See [164] for an account of the two separate histories of the development of public key cryptography (PKC).

Though the concept of PKE was introduced by Diffie and Hellman, they were unable to provide a concrete instantiation of such a scheme. It was left as an open problem until it was solved by three other researchers (Rivest, Shamir and Adleman [148]) and was called the RSA public key encryption system. Diffie and Hellman had introduced and solved another related and equally important problem. They considered the possibility of two parties performing some private computations and exchanging some messages over a public channel to finally arrive at a shared secret key. This is called public key agreement. The Diffie-Hellman key agreement (DH-KA) is shown in Figure 1.4.

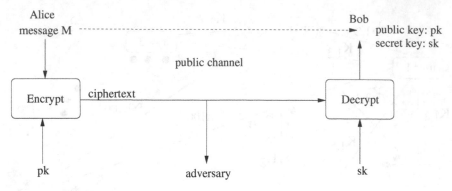

Fig. 1.3 An overview of public key encryption.

$$\text{Cyclic group } G = \{1, g, g^2, ..., g^{p-1}\}$$

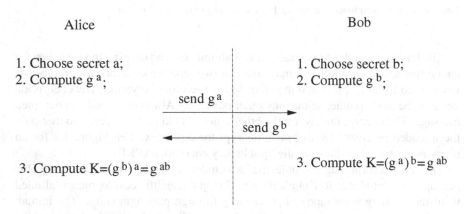

Fig. 1.4 Diffie-Hellman Public Key Agreement.

Note that the security of the Diffie-Hellman scheme relies on the facts that given g^a and g^b, no third party will be able to compute (i) one of the exponents a or b or (ii) the shared secret g^{ab}. The former is a classical problem in number theory called the discrete logarithm problem and the latter has become known as the Diffie-Hellman problem. So, the assurance that Alice and Bob get is conditioned on the fact that no one can solve these problems within a reasonable time. In fact, the security of all public key cryptographic schemes is based on the assumed hardness of some computational problem(s).

Later a public key encryption scheme was developed by ElGamal [79] which is very closely related to the Diffie-Hellman key agreement (DH-KA) protocol. This scheme is shown in Figure 1.5. This can be seen as a modification of the DH-KA. Bob does his part of the DH-KA protocol offline during the set-up phase and publishes (g, g^b) as his public key. Alice does her part of the DH-KA protocol online to compute g^a and the common secret key g^{ab}. This secret key is used to mask the message. Bob can recover the message by computing g^{ab} from g^a and his secret key b as in the DH-KA protocol.

It is indicative of the subtlety of the problem that Diffie and Hellman having proposed the notion of PKE and having discovered their key agreement protocol could not get the PKE scheme later proposed by ElGamal. It is also perhaps an issue of (a smaller scale) paradigm shift. In the ElGamal PKE scheme, the ciphertext consists of two group elements, but, the message is a single group element. So, there is a ciphertext expansion. In all previous encryption schemes, including the RSA scheme, the ciphertext is as long as the message. Though it is now considered routine, at that point of time in history, it was perhaps difficult to conceive a PKE scheme where the ciphertext is longer than the message.

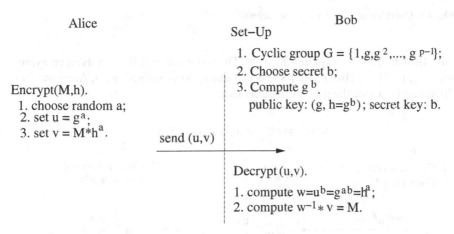

Fig. 1.5 ElGamal public key encryption scheme.

An important aspect of PKC is the notion of digital signatures. In this primitive, each user has a secret signing key and a public verification key. A message M is signed using the signing key to produce a signature σ. Anybody can verify the correctness of a message-signature pair using the public verification key. Concrete proposals of signature schemes were made using the RSA and the ElGamal PKE schemes. Figure 1.6 provides an overview of a digital signature scheme.

The advent of PKE is a major landmark in the evolution of cryptography. This, however, brought with it its own problems. Consider the situation shown in Figure 1.7. Here Eve is an active attacker, i.e., she can modify the information flowing across the public channel. So, she can intercept the messages that Alice and Bob exchange and replace them with messages of her own choosing. This allows her to establish secret keys separately with Alice and Bob without them realising this fact. In other words, Alice and Bob believe that they have established a shared key between themselves, whereas in actuality, they have established keys with Eve. So, if Alice sends a message to Bob using the key she believes that she has established

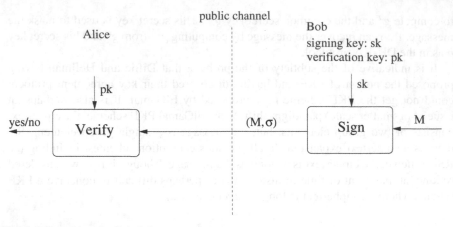

Fig. 1.6 Overview of a digital signature scheme.

with Bob, Eve can decrypt this message. The same is true if Bob sends an encrypted message to Alice. The crucial issue is that during key establishment, Alice and Bob do not authenticate themselves to each other.

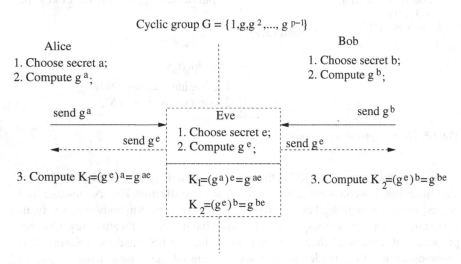

Fig. 1.7 (Wo)man in the middle attack.

PKE schemes are also subject to similar attacks. The issue is how to trust a public key? See Figure 1.8. When Bob wants to send a message to Alice, he will be using Alice's public key to encrypt the message. Suppose, Eve masquerades as Alice and puts forward a public key claiming it to belong to Alice. Eve, of course, knows the corresponding secret key. If Bob trusts this public key, then he will encrypt the message using this public key. Eve can decrypt the corresponding ciphertext thus, defeating the security of the system. To prevent this attack, Bob somehow needs to

be ensured that the public key that is claimed to belong to Alice indeed does belong
to Alice.

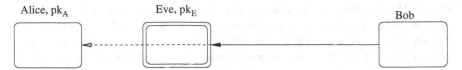

Fig. 1.8 How to trust a public key?

This is achieved using an authority that both Alice and Bob trust. This authority
issues certificates for public keys and is called a certifying authority (CA). To obtain
a certificate, Alice approaches the CA and submits her public key. The CA does the
necessary (physical) verification and determines the identity of Alice. After such
checking the CA uses a digital signature scheme to sign the public key of Alice.
When Bob wishes to send a message to Alice, he first obtains Alice's public key
and the certificate issued to Alice by the CA. Using the public verification key of
the CA, Bob can verify the signature of CA on Alice's public key. If this verification
succeeds, Bob trusts the public key to actually belong to Alice and uses it to encrypt
a message intended for Alice. This is pictorially shown in Figure 1.9.

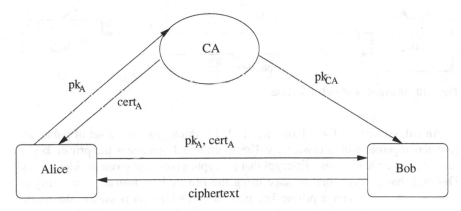

Fig. 1.9 Certifying authority and trust in public key.

Use of CA solves the problem of trust but, brings with it a host of other issues
regarding certificates. The basic question is the validity of a certificate. A certificate
might have been issued to Alice by the CA, but, this will have a time limit. Even
otherwise, Alice may have compromised her secret key and has to obtain a new public key/secret key pair and so, a new certificate on the new public key. The issue of
management of certificates is complex and cumbersome. There are no neat solutions
known and the issue is the main stumbling block in the widespread deployment of
PKE schemes.

1.2 Identity-Based Encryption

The problems associated with the practical deployment of PKE schemes motivated
Shamir [155] to introduce the concept of identity-based encryption (IBE). IBE is
a kind of public key encryption scheme where the public key of a user can be any
arbitrary string – typically the e-mail address. An overview of an IBE scheme is
given in Figure 1.10. When Bob wants to send a message to Alice; he encrypts it
using the e-mail id of Alice as the public key. There is no need for Bob to go to
the CA to verify the public key of Alice. This way an IBE can greatly simplify
certificate management. To quote Shamir [155]:

> *"It makes the cryptographic aspects of the communication almost transparent to the user,
> and it can be used effectively even by laymen who know nothing about keys or protocols."*

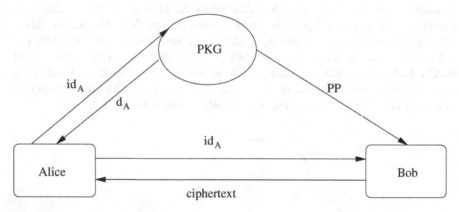

Fig. 1.10 An overview of an IBE scheme.

An IBE consists of four algorithms: Set-up which generates a set of public parameters together with a master key, Key-Gen which generates the private key of
an entity, given her identity, Encrypt that encrypts a message given the identity and
Decrypt that decrypts the message using the private key. Instead of a certifying
authority, here we have a private key generator (PKG) who possesses the master
secret key. In the above example, Bob authenticates himself to the PKG to obtain a
private key corresponding to his identity. (This can be even after he receives the encrypted message from Alice.) Bob uses this private key to decrypt all the messages
encrypted using his identity as the public key. Note that, Alice need not verify any
certificate relating the public key to send an encrypted message to Bob. What she
needs is the identity of Bob along with the public parameters of PKG.

Shamir posed a challenge to the crypto community to come out with a practical
IBE scheme. A satisfactory solution of this problem eluded cryptographers till the
turn of the millennium. The solution, when it finally arrived, came not from one,
but three different quarters – Boneh-Franklin [39], Sakai-Ohgishi-Kasahara [151]
and Cocks [70]. Among these, the former two based their cryptosystems on bilinear

pairing over elliptic curve groups while the last one was based on factoring and is less practical. Boneh and Franklin [39] formalised the notion of IBE, gave a precise definition of its security model and related the security of their construction with a natural analogue of the Diffie-Hellman problem in the bilinear setting, called the bilinear Diffie-Hellman (BDH) problem. This work caught the immediate attention of the crypto community world wide and turned out to be a major milestone within the paradigm of public key cryptography.

Boneh and Franklin use bilinear pairing over elliptic curve groups to construct their IBE. Initially bilinear maps such as Weil and Tate pairing were introduced in cryptology [133, 84] as weapons in the arsenal of the cryptanalyst – to reduce the discrete log problem in the elliptic curve group to the discrete log problem over finite fields. Joux [113] converted it into a tool of cryptography by proposing a one round tri-partite key agreement protocol using bilinear pairing. The works of Joux and Boneh-Franklin, in some sense, ignited an explosion in pairing based public key cryptography. Even a cursory glance at a compendium [17] of these works is enough to give a feel of the research activity that has been generated in the last one decade.

The concept of IBE was soon extended to hierarchical identity-based encryption (HIBE) [106, 94]. Requests for decryption keys are made to the PKG. A single PKG may become overloaded with such requests. The motivation for HIBE is to reduce the workload for the PKG. In a HIBE, instead of a single component, identities are considered to be vectors. Further, an entity having identity $id = (id_1, \ldots, id_j)$ and possessing the private key d_{id} can generate the private key of an entity whose identity is $id' = (id_1, \ldots, id_j, id_{j+1})$. This way HIBE reduces the workload of the PKG. The important thing to note is that the public parameters are that of the PKG, i.e., there are no lower level public parameters. Apart from being of some importance in its own right, a HIBE provides a flexible mechanism which can be used to construct other cryptographic primitives.

1.3 Plan of the Book

Technical discussion starts from the second chapter. Chapters 2 and 3 are of preliminary nature. Formal definitions of PKE and IBE schemes are given in Chapter 2. Extension to HIBE schemes are explained. The different variants of the security models for IBE and HIBE schemes are provided. Formal definition of IBE and an appropriate security model were proposed by Boneh and Franklin [39]. Later definitions and security models for related primitives were based on the foundation laid out in that work. An important aspect of the cryptographic schemes is the so-called security proofs or more accurately the security reductions. Chapter 2 discusses the overall structure of security proofs. This should help the reader in going through the proofs in the later chapters.

Most IBE schemes and certainly the practical ones are based on bilinear maps over elliptic curve groups. Discussing the relevant aspects of elliptic curves requires us to spend some space on finite field arithmetic. Chapter 3 provides the appropriate material on finite fields, elliptic curves and pairing. This material is not intended to be encyclopedic. Rather it is provided so that the reader can get some feel about how IBE schemes can actually be implemented.

The first publicly known IBE scheme is due to Boneh and Franklin [39]. The BF-IBE scheme is presented in Chapter 4 and the security reduction is discussed in some details. The proof assumes certain functions to be random functions – the so-called random oracle assumption. All schemes discussed in Chapter 4 have this common feature. Gentry and Silverberg [94] had extended the BF-IBE scheme to a HIBE scheme. Several variants of the BF-IBE scheme has been given by other authors. Chapter 4 provides a good idea of such schemes.

Historically, the first attempts to remove the random oracle assumption from the security proofs required a weakening of the security model. This is called the selective-identity model. Several elegant (H)IBE schemes have been proposed and shown to be secure in the selective-identity model. Chapter 5 provides a description of two important (H)IBE schemes. The first one is due to Boneh and Boyen [32] and the second one is due to Boneh, Boyen and Goh [35].

Design of IBE schemes can be thought of as a two-step procedure. In the first step, a scheme is designed which can be proved to be secure against chosen-plaintext attacks (CPA-secure). The proof technique for such schemes incorporate mechanisms to handle key extraction procedures. The second step is to attain security against chosen-ciphertext attacks (CCA-secure). Several methods have been proposed to convert CPA-secure schemes to CCA-secure schemes. Chapter 6 provides a discussion of these methods.

The next major step is to obtain IBE schemes which can be proved to be secure in the model introduced by Boneh and Franklin without using the random oracle assumption. An early construction of such a scheme was given by Boneh and Boyen [33]. This scheme, however, was quite impractical. An important variant of this scheme was given by Waters [169]. This variant is one of the most important IBE schemes proposed till date. Its importance stems both from the theoretical novelty of the construction as well as from its practicality. The only disadvantage of this scheme is the rather large size of its public parameters.

Independent work by Chatterjee and Sarkar [60] and Naccache [137] showed how to reduce the size of the public parameters with an associated degradation in the security bound. Chapter 7 discusses in details Waters IBE scheme and its variants. A new proof of Waters' IBE scheme was given by Bellare and Ristenpart [23]. The proof of Waters' IBE scheme that we provide is based upon the approach in [23]. Chapter 7 also provides a HIBE scheme secure against adaptive-identity attacks and its modification to attain CCA-security. This discussion is based on [61, 154].

One of the most important issues about the schemes in Chapter 7 is that the hardness assumption is the decisional bilinear Diffie-Hellman (DBDH) problem. This is the decision version of the bilinear Diffie-Hellman (BDH) problem introduced by Boneh and Franklin in [39]. The BDH and DBDH problems are regarded as the basic hardness assumptions in pairing based cryptography.

Subsequent important work on pairing based (H)IBE schemes have been done by Gentry [91] and Waters [170] (Note that this is different from Waters [169] discussed in Chapter 7). Gentry's work is important as it is the simplest scheme which can be proved to be secure against adaptive-identity attacks without the use of random oracles. The drawback is that the hardness assumption that is used is much more complex compared to the DBDH assumption. Waters [170] recently introduced a new technique called dual system encryption for designing IBE schemes. This technique, though relatively new, has been used in subsequent works and appears to hold out promise of further applications. Chapter 8 discusses Gentry's [91] and Waters' [170] IBE schemes and some of their implications.

Though most IBE schemes proposed till date use bilinear pairings, there has been some work on non-pairing based IBE schemes. These schemes are discussed in Chapter 9. The first such scheme is due to Cocks [70] and is based on the hardness of the quadratic residuacity problem. One problem with Cocks' scheme is that the size of the ciphertext is quite large. Boneh, Gentry and Hamburg [40] proposed an IBE scheme which reduces the size of the ciphertext (at the cost of increasing the time for encryption).

Lattice based techniques have recently found applications to the design of IBE schemes. The possibility of lattice based IBE was first discovered by Gentry, Peikert and Vaikuntanathan [93]. Later work using lattices have paralleled more or less the development of pairing based IBE schemes. Chapter 9 provides a somewhat detailed description of the lattice based method for designing IBE schemes.

IBE by itself is an important cryptographic primitive. Its further importance arises from the fact that both IBE and HIBE have proved to be useful in designing other cryptographic primitives and new functionalities. These include signatures, key agreement, broadcast encryption and public key encryption with keyword search. Some idea of such applications are given in Chapter 10.

A nagging issue with IBE is that of key escrow. The PKG knows the decryption key for every identity and consequently can decrypt any ciphertext formed using its public parameters. For real-life applications this can be a drawback. (In some situations, though, this might be exactly what is required.) Chapter 11 discusses the several approaches for dealing with the issue of key escrow in IBE schemes.

Though the idea of IBE is only about a decade old, some commercial products have already appeared in the market. Additionally some standards have been proposed. Finally, Chapter 12 briefly mentions such products and standards.

Chapter 2
Definitions and Notations

In this chapter, we present definitions and notation. We start with the definition of public key encryption schemes and their security models. This forms the basis of the corresponding notions for identity-based encryption schemes. The definition of IBE schemes is given and extended to that of HIBE schemes. Security model for HIBE schemes is defined. This security model can be specialised to that of IBE schemes by fixing the number of levels to one.

There are several variants of the security model for (H)IBE schemes. These are carefully explained and the notion of anonymity is defined. A related issue is the use of random oracles in the security analysis. We mention this briefly and discuss its relevance.

2.1 Public Key Encryption

A public key encryption (PKE) scheme is specified by three probabilistic algorithms. The run-time of each of these algorithms is upper bounded by a polynomial in a quantity called the security parameter, denoted by κ. This is formally expressed by explicitly providing 1^κ as input to the algorithms and requiring the run-times of the algorithms to be upper bounded by a polynomial in the length of this input. While this is formally appropriate, it is more convenient to simply note that the run-times are polynomially bounded in κ and avoid explicitly mentioning this.

Set-Up. This algorithm takes as input a security parameter κ. It outputs descriptions of the message space, the ciphertext space, the key space and a key pair (pk, sk) from the key space. Here pk is a public key and sk is the corresponding secret key. The pair (pk, sk) is randomly sampled from the key space. (Though it is not a definitional requirement, (pk, sk) would typically be uniformly distributed over the key space.)

Encrypt. It takes as input a message M and a public key pk and outputs a ciphertext C.

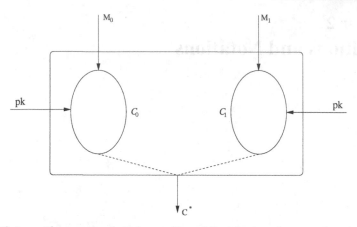

Fig. 2.1 $\mathcal{C}_0$ (resp. $\mathcal{C}_1$) corresponds to the set of possible ciphertexts that can arise when the encryption algorithm is applied to the message M_0 (resp. M_1). C^* is a uniform random choice from $\mathcal{C}_\gamma$, where γ is a uniform random bit.

Decrypt. It takes as input a ciphertext C and a private key sk and returns either a message M or the special symbol $\perp$. The symbol $\perp$ indicates that the ciphertext cannot be decrypted.

The encryption algorithm is a probabilistic algorithm and so there can be more than one ciphertext for a fixed message and a fixed public key. Equivalently, the encryption algorithm can be viewed as a sampling algorithm that given a message M and a public key pk samples from the set of possible ciphertexts which correspond to M and pk. Again the sampling will typically be done under the uniform distribution, though, it is not a definitional requirement.

A ciphertext can be said to be valid if it can be produced as an output of the encryption algorithm (on some pair of inputs M and pk) and invalid otherwise. The definition of the decryption algorithm does not require that the output has to be $\perp$ if the ciphertext is invalid; in this case, it may produce a random element of the message space as output.

For soundness, we require that if C is produced by Encrypt using pk, then the output of Decrypt on C using the corresponding secret key sk should give back M. Since the algorithms are probabilistic, the outputs are actually random variables over appropriate sets. In particular, the Set-Up algorithm can be seen to be sampling a pair of public and private keys from appropriate key spaces and the Encrypt algorithm samples from the set of possible ciphertexts which correspond to a message M and a public key pk. In principle, even though the Decrypt algorithm is allowed to be probabilistic, for most constructions, it is in fact a deterministic algorithm. We note that there are constructions, where the decryption algorithm is allowed to fail with an insignificant probability of error.

Next comes the question – how to define the security of a public key encryption scheme? A natural answer is – given a ciphertext no adversary should be able to learn any *meaningful* information about the corresponding plaintext. This intuitive notion is formalised into what is called *semantic security* in a landmark paper by Goldwasser and Micali [99]. They also provided a technical definition of security called *indistinguishabilty* and showed that for a passive attacker these two notions are equivalent. This result has later been extended to the case of an active adversary in [98, 168]. The equivalence between the natural notion of security and the technical definition turns out to be very important. Because it is more convenient to work with the technical definition of indistingushability than the natural notion of semantic security.

This technical notion of indistinguishability of ciphertexts for a PKE scheme in the case of a passive adversary can be easily understood with the help of Figure 2.1. For $i = 0, 1$, let $\mathscr{C}_i$ be the set of ciphertexts which may arise from the message M_i under the public key pk. The encryption algorithm defines a distribution over $\mathscr{C}_i$. Suppose that a bit γ is chosen uniformly at random and a ciphertext C^* is sampled from $\mathscr{C}_\gamma$ according to the distribution defined by the encryption algorithm.

An adversary is allowed to specify the messages M_0 and M_1; the bit γ is not revealed to the adversary, but, the ciphertext C^* is given to the adversary. Now the adversary has to guess the value of γ. If the adversary is unable to do so (with probability significantly away from half), then, to the adversary, the ciphertexts arising from M_0 are indistinguishable from the ciphertexts arising from M_1. This basic idea is built into an appropriate security model as we describe below for an active adversary.

Indistinguishability against adaptive chosen ciphertext attack [21] is the strongest accepted notion of security for a public key encryption scheme. An encryption scheme secure against such an attack is said to be IND-CCA2 secure. We give an informal description of IND-CCA2 security in terms of the following game between a challenger and an adversary $\mathscr{A}$, which is a probabilistic algorithm whose runtime is bounded above by a polynomial in the security parameter. Later we provide a more detailed explanation of the security game for an IBE scheme. Figure 2.2 gives an overview of the security game for a PKE scheme.

1. Given the security parameter κ, the challenger runs the Set-Up algorithm to generate a public and private key pair (pk, sk). It gives $\mathscr{A}$ the public key pk.
2. Given the public key, $\mathscr{A}$ *adaptively* issues decryption queries, which the challenger must properly answer. By adaptively it is meant that the adversary's next query can depend on the answers to the previous queries.
3. At some point, $\mathscr{A}$ outputs two equal length messages M_0, M_1 and the challenger responds with an encryption C^* of M_γ, where γ is a random bit.
4. The adversary continues with adaptive decryption queries but not on C^*.
5. Finally, $\mathscr{A}$ outputs its guess γ' of γ and wins if $\gamma' = \gamma$.

The advantage of $\mathscr{A}$ against the encryption scheme is

$$\mathsf{Adv}_{\mathscr{A}} = \left| \Pr[\gamma = \gamma'] - \frac{1}{2} \right|.$$

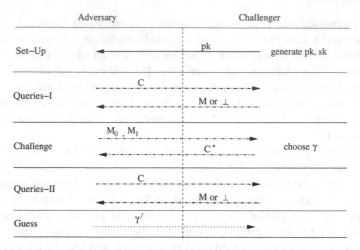

Fig. 2.2 A diagrammatic depiction of the five phases of the security model for a public key encryption scheme.

An encryption scheme is said to be (t,q,ε)-IND-CCA2 secure, if for all adversaries $\mathscr{A}$ running in time t and making at most q decryption queries, $\mathsf{Adv}_{\mathscr{A}} \leq \varepsilon$.

In case of a passive adversary, a weaker notion of security, called indistinguishability against chosen plaintext attack (in short IND-CPA security) of a public key encryption scheme is available in the literature [99, 21]. In the IND-CPA security game, the adversary is not allowed to place any decryption query. In other words, this is the scenario depicted in Figure 2.1 where the query phases depicted in Figure 2.2 are not allowed. Given a public key, the adversary simply outputs two equal length messages M_0, M_1 and the challenger responds with an encryption C^* of M_γ. The adversary wins if it can predict γ.

2.2 Identity-Based Encryption

The formal notion of an identity-based encryption scheme was developed in [155, 39]. An identity-based encryption scheme is specified by four probabilistic polynomial time (in the security parameter) algorithms: Set-Up, Key-Gen, Encrypt and Decrypt.

Set-Up: This algorithm takes as input a security parameter 1^κ, and returns the system parameters PP together with the master secret key msk. The system parameters include a description of the message space $\mathscr{M}$, the ciphertext space $\mathscr{C}$, the identity space $\mathscr{I}$ and the master public key. They are publicly known while the master secret key is known only to the private key generator (PKG). Usually, the descriptions of the different spaces are implicit in the description of the master public key and this itself is referred to as the public parameter PP.

Key-Gen: This algorithm takes as input an identity id $\in \mathscr{I}$ together with the public parameters PP and the master secret key msk and returns a private key d_{id}, using the master key. The identity id is used as the public key while d_{id} is the corresponding private key.

Encrypt: This algorithm takes as input an identity id $\in \mathscr{I}$, a message $M \in \mathscr{M}$ and the public parameters PP and produces as output a ciphertext $C \in \mathscr{C}$.

Decrypt: This takes as input a ciphertext $C \in \mathscr{C}$, an identity id, a corresponding private key d_{id} and the system parameters PP. It returns the message M or $\perp$ if the ciphertext cannot be decrypted.

These set of algorithms must satisfy the standard soundness requirement.

If
> (PP,msk) is output by Set-Up;
> d_{id} is a private key returned by Key-Gen for an identity id;
> C is a ciphertext produced by Encrypt on a message M,
> using identity id and public parameters PP;

then
> the output of Decrypt on C, id, d_{id} and PP should be M.

The comments regarding the encryption and decryption algorithms made in the context of PKE schemes are also applicable here. Additionally, similar comments apply to key generation. Given an identity and public parameters, it might be possible to have a set of corresponding decryption keys. In that case the key generation algorithm can be visualised as a strategy for sampling from this set. Note that the PKG can decrypt any message encrypted under any identity since it is the PKG who generated the private key for that identity. This is the so-called *key escrow* property of identity-based cryptography.

2.2.1 Hierarchical Identity-Based Encryption

Hierarchical identity-based encryption (HIBE) is an extension of IBE. The basic motivation for HIBE schemes is based on the following rationale. The generation of private key can be a computationally intensive task. The identity of an entity must be authenticated before issuing a private key and the private key needs to be transmitted securely to the concerned entity.

HIBE reduces the workload of the PKG by delegating the task of private key generation and hence authentication of identity and secure transmission of private key to its lower levels. However, only the PKG has a set of public parameters. The identities at different levels do not have any public parameters associated with them. Apart from being a standalone cryptographic primitive, HIBE has many interesting applications.

In contrast to IBE, for a HIBE identities are represented as vectors. So for a HIBE of maximum height h (which is denoted as h-HIBE) any identity id is a tuple

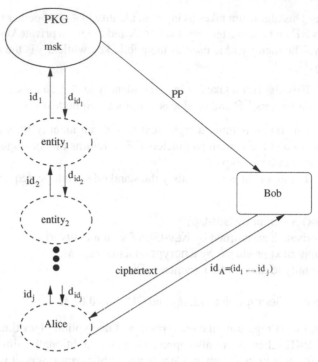

Fig. 2.3 Schematic diagram of the operation of a HIBE.

$(\mathsf{id}_1, \ldots, \mathsf{id}_\tau)$ where $1 \le \tau \le h$. Let, $\mathsf{id}' = \mathsf{id}'_1, \ldots, \mathsf{id}'_j$, $j \le \tau$ be another identity tuple. We say id' is a prefix of id if $\mathsf{id}'_i = \mathsf{id}_i$ for all $1 \le i \le j$.

As in the case of IBE, the PKG has a set of public parameters PP and a master key msk. For all identities at the first level the private key is generated by the PKG using msk. For identities at the second level onwards, the private key can be generated by the PKG or by any of the ancestors of that identity. In the above example, the private key d_{id} of id can be generated by an entity whose identity is a prefix of id and who has obtained the corresponding private key. This is shown in Figure 2.3.

The formal notion of a HIBE scheme is an extension of the corresponding notion of an IBE scheme and was developed in [106, 94]. A HIBE scheme $\mathscr{H}$ is specified by four probabilistic polynomial time (in the security parameter) algorithms: Set-Up, Key-Gen, Encrypt and Decrypt.

Set-Up: This algorithm takes input a security parameter 1^κ and returns the system (or public) parameters PP together with the master secret key msk. The system parameters include a description of the message space $\mathscr{M}$, the ciphertext space $\mathscr{C}$ and the identity space $\mathscr{I}$. These are publicly known while the master key is known only to the private key generator (PKG). In case, there is some maximal level h of the HIBE, then this is also made public.

Key-Gen: This algorithm takes as input an identity tuple $\mathsf{id} = (\mathsf{id}_1, \ldots, \mathsf{id}_j)$, $j \geq 1$ and the private key $d_{\mathsf{id}|j-1}$ for the identity $(\mathsf{id}_1, \ldots, \mathsf{id}_{j-1})$ and returns a private key d_{id} using $d_{\mathsf{id}|j-1}$. If $j = 1$, then $d_{\mathsf{id}|0}$ is defined to be the master secret key msk. The identity id is used as the public key while d_{id} is the corresponding private key.

Note that by appropriately invoking the **Key-Gen** algorithm the PKG as well as any proper predecessor of the identity tuple $(\mathsf{id}_1, \ldots, \mathsf{id}_j)$ can produce a decryption key for this identity.

Encrypt: This algorithm takes as input the system parameters PP, an identity id and a message M and produces as output a ciphertext C. This ciphertext is the encryption of M under the identity id and the public parameters PP.

Decrypt: This algorithm takes as input the public parameters PP, an identity id, a ciphertext C and a private key d_{id} and returns the message or $\perp$ if the ciphertext is not valid.

The standard soundness requirement that holds for IBE is also applicable for HIBE. If d_{id} is a private key corresponding to the identity tuple id generated by the **Key-Gen** algorithm and C is the output of the **Encrypt** algorithm for a message $M \in \mathcal{M}$ using id as a public key and PP; then the **Decrypt** algorithm must return M on input d_{id} and C.

Comments on the encryption and decryption algorithms made in the context of PKE schemes are applicable here. Further, comments made on the key generation algorithm in the context of IBE schemes are also applicable here. Also note that in addition to the PKG, any proper ancestor of id can decrypt messages encrypted under id.

2.3 Security Model for (H)IBE

As we have already noted, HIBE is a generalisation of IBE i.e., an IBE can be thought of as a single level HIBE. So instead of describing the security models for IBE and HIBE separately, we only describe the security model for HIBE. The security model for IBE is obtained by setting the number of levels to one.

The basic idea of the security model for (H)IBE schemes is obtained by extending the security model for PKE schemes. As in PKE we focus on the technical notion of indistingushability (it is known [13] that in the case of IBE also this technical notion is equivalent to the more natural notion of semantic security). Just like in PKE, it is a formalisation of the adversary's inability to distinguish between ciphertexts arising out of two equal length messages M_0 and M_1. An identity is chosen by the adversary as the target identity, i.e., the adversary's goal is to compromise the security of the identity it chooses as the target identity. A random bit γ is chosen and the challenge ciphertext is produced by encrypting M_γ under the target identity. The adversary wins if it can predict γ with a probability significantly away from half.

The main difference from PKE schemes is that a coalition of valid users of an IBE scheme can possibly launch an attack against another user of the scheme. Each valid user has a decryption key provided by the PKG for its identity. A group of users can form a coalition and attempt to compromise the security of another user. Modelling this aspect is a bit tricky, since the coalition may not be formed at the outset and may gradually grow. This modelling is done by providing the adversary with a key-extraction oracle. The adversary can query the oracle with an identity and receive a corresponding decryption key. It is allowed to place queries in an adaptive manner. Further, it can choose the target identity after making some key-extraction queries. In addition, as in PKE, the adversary may be given access to a decryption oracle. All of these is formalised in the following manner.

2.3.1 Chosen Ciphertext Attack

Recall that security against adaptive chosen ciphertext attack is the accepted notion of security for a public key encryption. This notion of security has been extended to the identity-based setting by Boneh and Franklin [39]. This is termed as IND-ID-CCA security (indistinguishability under adaptive identity and adaptive chosen ciphertext attack).

Let $\mathscr{H}$ be an h-HIBE scheme as defined in the previous section. The IND-ID-CCA security for $\mathscr{H}$ is defined [106, 94, 32] in terms of the following game between a challenger and an adversary of the HIBE. The adversary is allowed to place two types of oracle queries – decryption queries to a decryption oracle $\mathscr{O}_d$ and key-extraction queries to a key-extraction oracle $\mathscr{O}_k$. Figure 2.4 shows a schematic diagram of the security game defining the security of an IBE scheme. The notion of indistinguishability of ciphertexts is similar to the idea explained in the context of PKE schemes.

Set-Up. The challenger takes as input a security parameter 1^κ and runs the Set-Up algorithm of the HIBE. It provides $\mathscr{A}$ with the system parameters PP while keeping the master key msk to itself.

Phase 1: Adversary $\mathscr{A}$ makes a finite number of queries where each query is one of the following two types:

- key-extraction query (id): This query is placed to the key-extraction oracle $\mathscr{O}_k$. Questioned on id, $\mathscr{O}_k$ generates a private key d_{id} of id and returns it to $\mathscr{A}$.
  The Key-Gen algorithm is probabilistic and so if it is queried more than once on the same identity, then it may provide different (but valid) decryption keys. Some (H)IBE schemes can insist on storing the decryption key generated on the first query and returning the stored value on subsequent queries on the same identity. This can help in achieving a tight security reduction.
- decryption query (id, C): This query is placed to the decryption oracle $\mathscr{O}_d$. It returns the resulting plaintext or $\perp$ if the ciphertext cannot be decrypted.

$\mathscr{A}$ is allowed to make these queries adaptively, i.e., any query may depend on the previous queries as well as their answers.

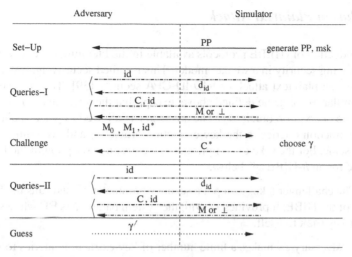

Fig. 2.4 A diagrammatic depiction of the five phases of the security model for identity-based encryption.

Challenge: When $\mathscr{A}$ decides that Phase 1 is complete, it fixes an identity id* and two equal length messages M_0, M_1 under the (obvious) constraint that it has not asked for the private key of id* or any prefix of id*. The challenger chooses uniformly at random a bit $\gamma \in \{0,1\}$ and obtains a ciphertext C^* corresponding to M_γ, i.e., C^* is the output of the Encrypt algorithm on input $(M_\gamma, \text{id}^*, \text{PP})$. It returns C^* as the challenge ciphertext to $\mathscr{A}$.

Phase 2: $\mathscr{A}$ now issues additional queries just like Phase 1, with the (obvious) restriction that it cannot place a decryption query for the decryption of C^* under id* or any of its prefixes nor a key-extraction query for the private key of id* or any prefix of id*. All other queries are valid and $\mathscr{A}$ can issue these queries adaptively just like Phase 1. The challenger responds as in Phase 1.

Guess: $\mathscr{A}$ outputs a guess γ' of γ.
The advantage of the adversary $\mathscr{A}$ in attacking the HIBE scheme $\mathscr{H}$ is defined as:

$$\text{Adv}_{\mathscr{A}}^{\mathscr{H}} = \left| \Pr[(\gamma = \gamma')] - 1/2 \right|.$$

An h-HIBE scheme $\mathscr{H}$ is said to be $(t, q_{\text{id}}, q_C, \varepsilon)$-secure against adaptive chosen ciphertext attack $((t, q_{\text{id}}, q_C, \varepsilon)$-IND-ID-CCA secure) if for any t-time adversary $\mathscr{A}$ that makes at most q_{id} private key queries and at most q_C decryption queries, $\text{Adv}_{\mathscr{A}}^{\mathscr{H}} \leq \varepsilon$. In short, we say $\mathscr{H}$ is IND-ID-CCA secure or when the context is clear, simply CCA-secure.

Shi and Waters [157] consider a more general security definition where the distribution of the keys depend on the actual delegation path. We do not consider this model in this work, since, for the schemes that we describe, the keys are uniformly distributed.

2.3.2 Chosen Plaintext Attack

Security reduction of (H)IBE protocols available in the literature generally concentrate on proving security in a weaker model. This is called security against adaptive identity chosen plaintext attack or IND-ID-CPA security [39]. The corresponding game is similar to the game defined above, except that the adversary is *not* allowed access to the decryption oracle $\mathcal{O}_d$. The adversary is allowed to place adaptive private key extraction queries to the key-extraction oracle $\mathcal{O}_k$ and everything else remains the same. For the sake of completeness, we give a description of the IND-ID-CPA game for an h-HIBE $\mathcal{H}$ below.

Set-Up The challenger takes as input a security parameter 1^κ and runs the Set-Up algorithm of the HIBE. It provides $\mathcal{A}$ with the system parameters PP while keeping the master key msk to itself.

Phase 1: Adversary $\mathcal{A}$ makes a finite number of key-extraction queries to $\mathcal{O}_k$. For a private key query corresponding to an identity id, the key-extraction oracle generates the private key d_{id} of id and returns it to $\mathcal{A}$. $\mathcal{A}$ is allowed to make these queries adaptively, i.e., any query may depend on the previous queries as well as their answers.

Challenge: At this stage $\mathcal{A}$ fixes an identity, id^* and two equal length messages M_0, M_1 under the (obvious) constraint that it has not asked for the private key of id^* or any of its prefixes. The challenger chooses uniformly at random a bit $\gamma \in \{0, 1\}$ and obtains a ciphertext (C^*) corresponding to M_γ, i.e., C^* is the output of the Encrypt algorithm on input $(M_\gamma, \mathrm{id}^*, \mathsf{PP})$. It returns C^* as the challenge ciphertext to $\mathcal{A}$.

Phase 2: $\mathcal{A}$ now issues additional queries just like Phase 1, with the (obvious) restriction that it cannot place a key-extraction query for the private key of id^* or any prefix of id^*. All other queries are valid and $\mathcal{A}$ can issue these queries adaptively just like Phase 1.

Guess: $\mathcal{A}$ outputs a guess γ' of γ.
Like the IND-ID-CCA game, the advantage of the adversary $\mathcal{A}$ in attacking the HIBE scheme $\mathcal{H}$ is defined as

$$\mathsf{Adv}_{\mathcal{A}}^{\mathcal{H}} = \left| \Pr[\gamma = \gamma'] - 1/2 \right|.$$

An h-HIBE scheme $\mathcal{H}$ is said to be (t, q, ε) secure against adaptive chosen plaintext attack if for any t-time adversary $\mathcal{A}$ that makes at most q private key extraction queries, $\mathsf{Adv}_{\mathcal{A}}^{\mathcal{H}} \leq \varepsilon$. In short we say $\mathcal{H}$ is (t, q, ε)-IND-ID-CPA secure or simply CPA-secure if the context is clear.

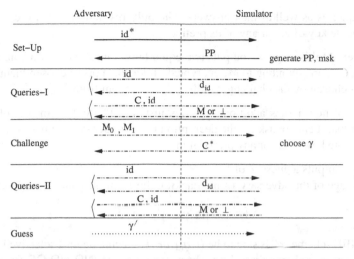

Fig. 2.5 A diagrammatic depiction of the five phases of the selective-identity security model for identity-based encryption.

2.3.3 Selective-ID Model

A weaker definition of security for identity-based encryption schemes is the so called *selective*-ID model [53, 54]. In this model, the adversary $\mathscr{A}$ commits to a target identity before the system is set up. This notion of security is called the selective identity, chosen ciphertext security (IND-sID-CCA security in short).

Compared to the security model where the adversary can choose the target identity adaptively, this is a very restricted notion of security. Correspondingly, it is also significantly easier to argue security in this model. If the sole interest is in obtaining a secure IBE, then selective identity security model is not satisfactory. On the other hand, this model is useful in other ways. In particular, it provides a new method to convert CPA-secure IBE schemes to CCA-secure PKE schemes.

Following [53, 54, 32] we define IND-sID-CCA security for an h-HIBE in terms of the game described below. A schematic diagram for the selective-identity security model is shown in Figure 2.5.

Initialization: The adversary outputs a target identity tuple $\mathsf{id}^* = (\mathsf{id}_1^*, \ldots, \mathsf{id}_u^*)$, $1 \leq u \leq h$ on which it wishes to be challenged.

Set-Up: The challenger sets up the HIBE and provides the adversary with the system public parameters PP. It keeps the master key msk to itself.

Phase 1: Adversary $\mathscr{A}$ makes a finite number of queries where each query is either a decryption or a key-extraction query. In a decryption query, it provides the ciphertext as well as the identity under which it wants the decryption. Similarly, in a key-extraction query, it asks for the private key of the identity it provides. Further, $\mathscr{A}$ is allowed to make these queries adaptively, i.e., any query may depend on the previous queries as well as their answers. The only restriction is that it cannot ask for the private key of id* or any of its prefixes.

Challenge: At this stage, $\mathscr{A}$ outputs two equal length messages M_0, M_1 and gets a ciphertext C^* corresponding to M_γ encrypted under the a priori chosen identity id*, where γ is chosen by the challenger uniformly at random from $\{0, 1\}$.

Phase 2: $\mathscr{A}$ now issues additional queries just like Phase 1, with the (obvious) restriction that it cannot ask for the decryption of C^* under id* or any of its prefixes nor the private key of id* or any prefix of id*.

Guess: $\mathscr{A}$ outputs a guess γ' of γ.
The advantage of the adversary $\mathscr{A}$ in attacking the HIBE scheme $\mathscr{H}$ is defined as:

$$\mathsf{Adv}_{\mathscr{A}}^{\mathscr{H}} = \left| \Pr[\gamma = \gamma'] - 1/2 \right|.$$

The HIBE scheme $\mathscr{H}$ is said to be $(t, q_{\mathsf{id}}, q_C, \varepsilon)$-secure against selective identity, adaptive chosen ciphertext attack (in short, $(t, q_{\mathsf{id}}, q_C, \varepsilon)$-IND-sID-CCA secure) if for any t-time adversary $\mathscr{A}$ that makes at most q_{id} private key extraction queries and at most q_C decryption queries, $\mathsf{Adv}_{\mathscr{A}}^{\mathscr{H}} \leq \varepsilon$.

Note that, in the above game the adversary has to commit to an identity tuple even before the appropriate spaces are defined by the Set-Up algorithm.

We may restrict the adversary from making any decryption query. An h-HIBE scheme $\mathscr{H}$ is said to be $(t, q_{\mathsf{id}}, \varepsilon)$-secure against selective identity, adaptive chosen plaintext attack (in short, $(t, q_{\mathsf{id}}, \varepsilon)$-IND-sID-CPA secure) if for any t-time adversary $\mathscr{A}$ that makes at most q_{id} private key queries, $\mathsf{Adv}_{\mathscr{A}}^{\mathscr{H}} \leq \varepsilon$.

2.3.4 Anonymous (H)IBE

There is another aspect which is sometimes added to the security definition of (H)IBE. This concerns the anonymity of the ciphertext. Informally, the idea is as follows. For certain applications, it may be a security concern that the identity of the intended recipient is revealed from the ciphertext. In other words, looking only at C (and PP) it may be possible to determine the identity id which has been used to generate C from some message. Roughly speaking, if for a (H)IBE scheme this is not possible, then the scheme is said to be anonymous. The formal definition of anonymity is obtained by modifying the above security game.

1. The Set-Up and the query phases 1 and 2 of the security game remain unchanged.
2. In the challenge stage, the adversary submits two identities id_0 and id_1 along with two equal length messages M_0 and M_1.
3. The simulator chooses two independent and uniform random bits γ_1 and γ_2 and provides the adversary with C^* which is the encryption of M_{γ_1} to the identity id_{γ_2}.

4. In the guess stage, the adversary outputs two bits γ_1' and γ_2' and wins if $\gamma_1 = \gamma_1'$ and $\gamma_2 = \gamma_2'$.
5. The adversary's advantage is defined to be

$$\left| \Pr\left[(\gamma_1 = \gamma_1') \wedge (\gamma_2 = \gamma_2') \right] - \frac{1}{4} \right|.$$

Parameterization of an adversary is done in the manner it is done for the (H)IBE schemes without anonymity considerations.

In any security game for (H)IBE schemes, the key-extraction oracle is always provided to the adversary. Additionally, the different variants of the security game are based on the following three points.

1. Whether the decryption oracle is provided to the adversary or not.
2. Whether the adversary can choose the target identity adaptively, i.e., after making queries to the key-extraction oracle; or whether the adversary has to commit to the target identity before the scheme is set-up giving the selective-identity game.
3. Whether the (H)IBE scheme is defined for the anonymity game or not.

The convention is that if anonymity is required, then it is specifically mentioned. For the other two points, there are four options leading to four different sets of restrictions on the adversary. Among these, the most restricted adversary is selective-identity and not having access to the decryption oracle which leads to the weakest (among the above four) notion of security for a (H)IBE scheme. The most powerful adversary is adaptive-identity and can make decryption queries leading to the strongest notion of security for a (H)IBE scheme. Several other security notions such as one-wayness, non-malleability, semantic security and multiple-target multiple-challenge CCA security have been formulated in the context of IBE and it is shown [13] that IND-ID-CCA security implies all other notions of security. In this sense IND-ID-CCA security can be considered as the "right" notion of security in the identity-based setting.

As we will discuss later, there are generic methods to convert a CPA-secure (H)IBE scheme to a CCA-secure (H)IBE scheme. Non-generic methods which apply to a fairly large number of schemes are also known. Due to these results, almost all schemes start out by first obtaining CPA-security. Among CPA-secure schemes, the issue is whether the scheme satisfies selective-identity or adaptive identity security. We will be seeing both kinds of schemes later.

2.3.5 Use of Random Oracles

The security analyses of (H)IBE schemes essentially show that if an adversary can win a security game (one among the variants described above), then it is possible to use such an adversary to solve some problem which is conjectured to be computationally hard. This is the usual reductionist technique used in complexity theory. We will be seeing some of these conjectured hard problems later. For the moment, we would like to discuss a different issue.

During the security reduction (or proof), it may be necessary to assume that certain functions described in the scheme are uniform random functions. In other words, if a set of distinct inputs are provided to such a function, then the outputs are independent and uniformly distributed. Clearly, such an assumption cannot hold for any practical function that may be used to instantiate the scheme. Alternatively, this can be viewed as an idealisation of some functions that are actually used in practice.

This is the so-called "random oracle" assumption used to argue about the security of many cryptographic primitives and not only (H)IBE schemes. This technique was formally introduced in [24] and has been criticised later in [52]. A rebuttal of the criticism and a robust defence of the random oracle technique has been given in [124].

Clearly, it would be better if one could build schemes where one does not have to use random oracles and there are indeed such schemes. At the same time, one should note that schemes for which the proofs require "random oracles" may also be useful in two situations: when no other scheme is known which does not require "random oracle proofs" (under the same kind of hardness assumption) and when the use of random oracles improves efficiency.

2.4 Structure of Security Proofs

We provide a few words on the structure of security proofs. All proofs are reductions. Suppose a cryptographic protocol is built upon several other smaller protocols. Then the assurance provided by a reduction is of the following form.

If

(smaller protocols are secure
and)
some problem Π is computationally hard

then

the main protocol is secure.

The argument is established through a contradiction. One starts with the assumption that there is an adversary who can break the main protocol with some non-negligible advantage in the given security model. This adversary is then used as a blackbox to construct an algorithm that either solves the underlying hard computational problem Π or breaks one of the smaller protocols with non-negligible probability of success. This contradicts the original hypothesis. As mentioned earlier, the proof itself may assume some hash functions to be uniform random functions.

A convenient way to structure the reductionist proof is to consider a sequence of games [160, 26]. To prove, for example, the indistinguishability of the encryption of two equal length plaintexts we construct a sequence of games of the following form.

A Game Sequence

$G_0,$

$G_1,$

$\vdots$

G_k

- Let X_i be the event that $\gamma = \gamma'$ in Game G_i. We consider

$$Pr[X_0],$$
$$Pr[X_0] - Pr[X_1],$$
$$\vdots$$
$$Pr[X_{k-1}] - Pr[X_k]$$
$$Pr[X_k].$$

In the above sequence, the following points are to be noted

1. G_0 is the game which defines the security of the protocol and so

$$\mathsf{Adv}(\mathscr{A}) = |Pr[\gamma = \gamma'] - 1/2| = |Pr[X_0] - 1/2|.$$

2. G_k is designed such that the bit γ is statistically hidden from the adversary. So,

$$Pr[X_k] = 1/2.$$

3. Games G_{i-1} and G_i differ:

   a. the difference is not too much;
   b. the adversary should not be able to notice whether he is playing Game G_{i-1} or Game G_i.

4. More precisely, $Pr[X_{l-1}] - Pr[X_i]$ is bounded above by

   a. either, the advantage of an adversary in breaking one of the smaller protocols;
   b. or, the advantage of solving problem Π.

$$\begin{aligned}
\mathsf{Adv}(\mathscr{A}) &= |Pr[X_0] - 1/2| \\
&= |Pr[X_0] - Pr[X_k]| \\
&\leq |Pr[X_0] - Pr[X_1]| \\
&\quad + |Pr[X_1] - Pr[X_2]| \\
&\quad + \cdots \\
&\quad + |Pr[X_{k-1}] - Pr[X_k]|.
\end{aligned}$$

If the adversary has a non-negligible advantage then there must be at least two consecutive games, X_{i-1} and X_i such that $|\Pr[X_{i-1}] - \Pr[X_i]|$ is non-negligible – which contradicts the original hypothesis.

2.5 Conclusion

This chapter introduced the necessary formalism. Later chapters describing different constructions of IBE schemes will be based on the formal notions introduced here. Apart from this minimal requirement, this chapter also provided background intuition behind the various definitions. This intuition will be useful in working through the proofs and constructions given later.

Chapter 3
A Brief Background on Elliptic Curves and Pairings

A bilinear pairing suitable for use in cryptography is a non-degenerate, efficient-to-compute, bilinear map

$$e : G_1 \times G_2 \to G_T$$

where G_1, G_2 and G_T are cyclic groups of the same prime order p. Known examples of such maps arise from certain algebraic geometric objects. The most commonly used pairings arise from the theory of elliptic curves where G_1 and G_2 are subgroups of points on an elliptic curve over a finite field whereas G_T is a subgroup of the multiplicative group of a finite field. Customarily, the group operation for elliptic curves is denoted by an addition and so we write G_1 and G_2 additively whereas we write G_T multiplicatively.

The purpose of this chapter is to provide a brief idea of how such maps are defined. This discussion is by no means exhaustive and the different techniques used to speed-up pairing computations is really outside the scope of this book. The other aspect of pairings is one of associated hard computational problems. For reduction style proofs of security of different schemes, one requires to assume that certain problems are computationally hard. We present a careful description of some of the important problems that have been used in the literature. Again, we do not aim to provide an exhaustive list of hard computational problems over pairing groups.

Before getting into details, we briefly discuss what is meant by non-degenerate, efficient and bilinear properties of the map e. Let $G_1 = \langle P_1 \rangle$ and $G_2 = \langle P_2 \rangle$. Non-degeneracy means that if $e(P, Q)$ is the identity element of G_T, then either P is the identity of G_1 or Q is the identity of G_2.

Efficiency is taken to mean that there is a polynomial time algorithm which can compute the map e. Typically, the logarithm (to the base two) of the sizes of G_1, G_2 and G_T will be bounded above by a polynomial of a quantity called the security parameter. So, the representations of the elements of G_1, G_2 and G_T will be bounded above by a polynomial in the security parameter and which in turn implies that the run-time of the algorithm to compute e will also be bounded above by a polynomial in the security parameter. While this is the asymptotic notion of efficiency, for practical implementations, this is not enough. The requirement is to have algorithms to compute pairings so that commercial deployment of the associated schemes becomes viable. This has motivated tremendous amount of research into finding efficient algorithms for pairing computations which as mentioned above is outside the scope of this book. Interested readers are referred to [104, 128, 167] for more details.

Bilinearity means that the map e is linear in both components which refers to the following two properties.

- $e(R_1 + R_2, Q) = e(R_1, Q)e(R_2, Q)$;
- $e(R, Q_1 + Q_2) = e(R, Q_1)e(R, Q_2)$.

A consequence of these two properties is the following fact which forms the basic algebraic novelty upon which all bilinear pairing based schemes are based. For $a, b \in \mathbb{Z}_p$.

$$e(aP_1, bP_2) = e(bP_1, aP_2) = e(P_1, P_2)^{ab}.$$

The other requirement on the map e is that certain computational problems associated with the elements of G_1, G_2 and G_T must be hard. Here hardness refers to the fact that at present there are no *known* probabilistic polynomial time algorithms to solve them. The general belief in the community is that such problems indeed cannot be solved in probabilistic polynomial time, but, currently, we are far from proving any such statement. Later in this chapter, we discuss some of the basic hard problems that have been proposed in the literature.

3.1 Finite Fields, Elliptic Curves and Tate Pairing

In this section, we provide an overview of some facts regarding the algebraic structures required to construct and compute a bilinear map. We start with the basic idea of exponentiation in a general cyclic group, briefly discuss finite fields and elliptic curves and present a definition of Tate pairing and describe Miller's algorithm for computing it. For more details on finite fields the reader may refer to [132] while for implementation of finite field and elliptic curve operations one may refer to [103]. Details of pairing operations can be found in [75]. The online source [28] is an excellent collection of explicit formulae for different operations over elliptic curves and pairing computations.

3.1.1 Exponentiation in General Cyclic Groups

Let $\langle g \rangle$ be a (multiplicatively written) cyclic group of order q. Exponentiation in $\langle g \rangle$ refers to be the following computation: given $a \in \mathbb{Z}_q$ compute $h = g^a$.

Let $a = a_{n-1} \ldots a_0$, where $n = \lceil \log_2 q \rceil$ and each a_i is a bit. There are two simple square-and-multiply methods for performing this computation – one which works from right to left, i.e. from a_0 to a_{n-1} and the other works from left to right, i.e., from a_{n-1} to a_0.

The right-to-left method can be explained as follows.

- $n = 1: h = g^{a_0}$.
- $n = 2: h = g^{2a_1 + a_0} = (g^2)^{a_1} \times g^{a_0}$.
  $t = g; r = g^{a_0};$
  $t = t^2; r = t^{a_1} \times r; h = r.$
- At ith step: square t; multiply t to r if $a_i = 1$.

Similarly, the left-to-right method can be explained as follows.

- $n = 1: h = g^{a_0}$.
- $n = 2: h = g^{2a_1 + a_0} = (g^{a_1})^2 \times g^{a_0}$.
  $r = g^{a_1};$
  $r = r^2 \times g^{a_0}; h = r.$
- At ith step: square r; multiply r by g if $a_{n-i} = 1$.

The importance of the left-to-right method arises from the fact that multiplication is always done by g which is a constant throughout the whole computation. In the right-to-left method, both the multiplicands change as the computation proceeds. Multiplying with a fixed element can provide certain efficiency benefits in implementation.

Both the above square-and-multiply algorithms are special cases of a strategy called addition chain method. An addition chain of length ℓ is a sequence of $\ell + 1$ integers such that the first integer is 1; each subsequent integer is a sum of two previous integers. For example, the following sequence is an addition chain: 1,2,3,5,7,14,28,56,63 ($2 = 1 + 1$, $3 = 2 + 1$, $5 = 3 + 2$, $7 = 5 + 2$, $14 = 7 + 7$, $28 = 14 + 14$, $56 = 28 + 28$, $63 = 56 + 7$). Clearly, addition chains can be used to compute exponentiation. However, finding a short addition chain is a computationally hard problem. An excellent survey of exponentiation algorithms has been made by Bernstein [27].

The discrete logarithm problem. The discrete log problem over a cyclic group $\langle g \rangle$ is the following. Given g and $h \in \langle g \rangle$, find a such that $h = g^a$. This a is called the discrete log of h with respect to base g and is written as $\log_g h = a$. Suppose n is the logarithm to base two of the size of $\langle g \rangle$. For a cyclic group to be useful in cryptography, the discrete log problem over the group should be computationally hard. By this we mean that there should no algorithm which runs in time polynomial in n. Clearly, one can find a by checking all the elements of $\langle g \rangle$, but, this requires time $O(2^n)$ which is exponential in n. Generic algorithms such as the Pollard's rho algorithm for finding discrete log requires $O(2^{n/2})$ time, which is better than $O(2^n)$ but is still exponential. The discrete log problem over certain groups obtained from finite fields and elliptic curves are believed to be computationally hard and are used to implement cryptographic operations.

We mention two other important computational problems. In the computational Diffie-Hellman (CDH) problem, the instance consists of a triplet (g, g^a, g^b) and the requirement is to compute g^{ab}. The decisional Diffie-Hellman (DDH) problem is to distinguish between the distributions (g, g^a, g^b, g^{ab}) and (g, g^a, g^b, g^c) for independent and uniform random a, b and c. We provide more formal definitions later.

3.1.2 Finite Fields

Let $(\mathbb{F}, +, \cdot)$ be a finite field so that $(\mathbb{F}, +)$ is a commutative group, $\mathbb{F}^* = (\mathbb{F} \setminus \{0\}, \cdot)$ is a cyclic group and multiplication distributes over addition. The number of elements in $\mathbb{F}$, denoted as $\#\mathbb{F}$, equals p^m, where p is a prime and $m \geq 1$. The prime p is called the characteristic of the field. All fields of the same cardinality are isomorphic and one customarily writes $\mathbb{F}_{p^m}$ to denote the finite field having p^m elements. The basic operations on a finite field are addition (and subtraction), multiplication and inversion (or division). Efficiencies of these operations depend on the actual representation of the field.

The fields that are useful in cryptography are "large" fields with $p^m \geq 2^{512}$. Some commonly used fields are of the following types:

large characteristic: $m = 1$ and p is "large";
characteristic 2: $p = 2$;
characteristic 3: $p = 3$.

Characteristic 3 fields have been suggested mainly for pairing based cryptography. Other composite order fields such as optimal extension fields [15] have been suggested for cryptographic use. The basic criteria for choosing a field is the security/efficiency trade-off which we briefly discuss below.

If p is a large prime, then one requires efficient multi-precision arithmetic to implement the operations of $\mathbb{F}_p$. The elements of $\mathbb{F}_p$ are stored as machine-sized words and computations are done modulo p. Multiplication is implemented using a combination of Karatsuba's algorithm and table look-up whereas inversion is done using Itoh and Tsuji's method [109]. The time for an inversion is substantially more than the time for a multiplication.

If p is two, then implementation is done using the so-called polynomial basis. Let $\tau(x)$ be an irreducible polynomial of degree m over $\mathbb{F}_2$. Then $\mathbb{F}$ is realised as the set of all polynomials of degree at most $m - 1$ over $\mathbb{F}_2$. Addition and multiplication are done modulo $\tau(x)$. Efficient implementation of multiplication can still be done using Karatsuba's algorithm whereas inversion can be done using the extended Euclidean algorithm. In this case, the ratio of times for an inversion to a multiplication is lower than that for large characteristic prime. There is another way to represent the field elements using a so-called normal basis. In this representation, squaring is 'free' but multiplication is costlier.

For cryptographic purposes, the whole of $\mathbb{F}_{p^m}^*$ is not used. Let r be a prime dividing $p^m - 1$. Then $\mathbb{F}_{p^m}^*$ has a subgroup of order r. Being of prime order, this subgroup is cyclic, i.e., $G = \langle g \rangle$. Cryptography is done over this subgroup. The basic necessary criteria is that the discrete log problem should be hard. Some basic methods for computing discrete log over finite fields are the following.

1. Pollard's rho method requires $O(\sqrt{r})$ time.
2. The index calculus method requires time $O\left(e^{(1+o(1))\sqrt{\ln p \ln \ln p}}\right)$ and works over $\mathbb{Z}_p^*$.
3. The number field sieve [129] is the best known method for finding discrete log over finite fields. The run time of this method is

$$O\left(e^{(1.92+o(1))(\ln p^m)^{1/3}(\ln \ln p^m)^{2/3}}\right).$$

This is a sub-exponential expression, i.e., it is neither fully polynomial nor fully exponential. This time complexity of this algorithm is determined by the size p^m of the entire field and not by the size r of the subgroup in which cryptography is performed.

The basic security/efficiency trade-off is determined by the above considerations. The value of r and the size of $\mathbb{F}$ has to be chosen so that all known algorithms require at least a certain amount of time to solve the discrete logarithm problem. The size of the field determines the size of representation of an element of the field which in turn, determines the efficiency of the field operations. The existence of sub-exponential algorithms necessitates that the size p^m of the field should be quite large. A detailed study of feasible parameters has been made by Lenstra and Verheul [130].

3.1.3 Elliptic Curves

The theory of elliptic curves is quite old. Application to cryptography was independently proposed by Koblitz [122] and Miller [134]. Since then a lot of work has been done on this topic. Below we provide a brief background on elliptic curves. See [163] for a standard introduction to the theory of elliptic curves.

Let K be a field. An elliptic curve in Weierstraß form over K is given by the following equation.

$$E/K : y^2 + a_1 xy + a_3 y = x^3 + a_2 x^2 + a_4 x + a_6, \qquad (3.1.1)$$

where $a_i \in K$ and there are no "singular points" and there is one rational point $\mathcal{O}$ called the point at infinity. Suppose L is an extension field of K. Then the set of L-rational points on E is defined to be the set

$$E(L) = \{(x,y) \in L \times L : y^2 + a_1 xy + a_3 y = x^3 + a_2 x^2 + a_4 x + a_6 = 0\} \cup \{\mathcal{O}\}.$$

If $L \supseteq K$, then $E(L) \supseteq E(K)$. The algebraic closure of K is denoted by $\overline{K}$ and E denotes the set $E(\overline{K})$.

If the characteristic of K is not equal to 2 or 3, then the above form can be simplified as follows. Replacing y by $\frac{1}{2}(y - a_1 x - a_3)$ gives

$$y^2 = 4x^3 + b_2 x^2 + 2b_4 x + b_6$$

where $b_2 = a_1^2 + 4a_2, b_4 = 2a_4 + a_1a_3, b_6 = a_3^2 + 4a_6$. Replacing (x,y) by $((x - 3b_2)/36, y/108)$ gives

$$y^2 = x^3 - 27c_4 x - 54c_6.$$

Define

$$b_8 = a_1^2 a_6 + 4a_2 a_6 - a_1 a_3 a_4 + a_2 a_3^2 - a_4^2$$
$$c_4 = b_2^2 - 24b_4$$
$$c_6 = -b_2^3 + 36b_2 b_4 - 216b_6$$
$$\Delta = -b_2^2 b_8 - 8b_4^3 - 27b_6^2 + 9b_2 b_4 b_6$$
$$j = c_4^3/\Delta$$
$$\omega = dx/(2y + a_1 x + a_3)$$
$$= dy/(3x^2 + 2a_2 x + a_4 - a_1 y).$$

Δ is called the discriminant, j is called the j-invariant and ω is called the invariant differential. The following relations hold: $4b_8 = b_2 b_6 - b_4^2$ and $1728\Delta = c_4^3 - c_6^2$.

So, if the characteristic of K is not equal to 2 or 3, then Equation (3.1.1) simplifies to the form

$$y^2 = x^3 + ax + b \qquad (3.1.2)$$

where $a, b \in K$ and $4a^3 + 27b^2 \neq 0$. The last condition ensures that $x^3 + ax + b$ does not have repeated roots. ($x^3 + ax + b$ has repeated roots if and only if $x^3 + ax + b$ and $\frac{d}{dx}(x^3 + ax + b) = 3x^2 + a$ have a common root; eliminating x from these two relations gives the condition $4a^3 + 27b^2 = 0$ and this corresponds to $\Delta = 0$.)

If the characteristic of K is equal to 2 then the Equation (3.1.1) can be simplified to [123]

- $y^2 + xy = x^3 + ax^2 + b$, $a, b \in K$, $b \neq 0$ (non-supersingular), or
- $y^2 + cy = x^3 + ax + b$, $a, b, c \in K$, $c \neq 0$ (supersingular).

Later, we explain what is meant by a supersingular elliptic curve.

Group law. The set of points E together with a special point at infinity (denoted by $\mathcal{O}$) can be made into a group using a suitably defined group law. The group law is written additively and is geometrically defined using the so-called chord-and-tangent rule. Suppose P and Q are any two points on E given by (3.1.2). The basic rules are as follows.

- $P + \mathcal{O} = \mathcal{O} + P = P$.
- $-\mathcal{O} = \mathcal{O}$.
- If $P = (x, y)$, then $-P = (x, -y)$.
- If $Q = -P$, then $P + Q = \mathcal{O}$.

Suppose now that $P = (x_1, y_1), Q = (x_2, y_2)$ and $P \neq -Q$. If $P \neq Q$, then the line $\ell(x,y) : y = \lambda x + v$ through P and Q intersects the curve E at a third point R; the reflection of R on the x-axis is defined to be the point $P + Q$ given by (x_3, y_3). If

$P = Q$, then the tangent $\ell(x,y) : y = \lambda x + v$ intersects the curve at a point R; the reflection of R on the x-axis is defined to be the point $2P$ given by (x_3, y_3).

If $P \neq Q, -Q$, then $\lambda = (y_2 - y_1)/(x_2 - x_1), v = y_1 - \lambda x_1 = y_2 - \lambda x_2$. Putting $\ell(x,y)$ into the equation of the curve we get $(\lambda x + v)^2 = x^3 + ax + b$ which is the same as $x^3 - \lambda^2 x^2 + (a - 2v\lambda)x + b - v^2 = 0$. This equation has three roots and x_1, x_2 are two of the roots. So the third root is $x_3 = \lambda^2 - x_1 - x_2$. Also, $-y_3 = \lambda x_3 + v$ and $y_1 = \lambda x_1 + v$ gives $y_3 = \lambda(x_1 - x_3) - y_1$. (Note that the line through (x_1, y_1) and (x_2, y_2) passes through $(x_3, -y_3)$.) Now suppose that $P = Q$. Using the equation of the curve $y^2 = x^3 + ax + b$ we have, $2y\frac{dy}{dx} = 3x^2 + a$ and so the slope λ at (x_1, y_1) is $\frac{3x_1^2 + a}{2y_1}$. The rest of the analysis is the same as the previous case. The obtained formula for (x_3, y_3) is same except for the changed value of λ. Algebraically these two cases are computed as follows.

If $P = (x_1, y_1)$, $Q = (x_2, y_2)$, with $P \neq -Q$, then $P + Q = (x_3, y_3)$, where $x_3 = \lambda^2 - x_1 - x_2$ and $y_3 = \lambda(x_1 - x_3) - y_1$, with

$$\lambda = \begin{cases} = \frac{y_2 - y_1}{x_2 - x_1} & \text{if } P \neq Q; \\ \\ = \frac{3x_1^2 + a}{2y_1} & \text{if } P = Q. \end{cases}$$

The group axioms can be directly verified using the above addition law, the only difficult case being the associative rule. The associative rule can be better seen from a more sophisticated algebraic approach using the notion of divisors. Later we will briefly mention divisors.

If K is a finite field and L is a finite extension of K, then the set $E(L)$ of the L-rational points of E form a finite subgroup of the group E. Cryptography is done over a suitable prime order subgroup of $E(L)$.

Some properties of elliptic curves. An important map in the context of elliptic curve defined over a finite field is the *Frobenius* map.

$$\tau_p : E(\overline{\mathbb{F}}_p) \to E(\overline{\mathbb{F}}_p), \qquad \tau_p : (x,y) \mapsto (x^p, y^p).$$

The map τ_p is a group homomorphism and the *trace of Frobenius* is defined to be $t_p = p + 1 - \#E(\mathbb{F}_p)$.

If L is a finite extension of a finite field K, then the number of points in $E(L)$ is given by *Hasse's theorem* to be $\#E(L) = \#L + 1 - t$, where $|t| \leq 2\sqrt{(\#L)}$. Consequently, $\#E(L) \approx \#L$. *Weil's theorem* relates the number of points in $E(K)$ to the number of points in $E(L)$ where L is a degree k extension of K with $\#K = q$. Let $t = q + 1 - \#E(K)$ and α, β be the complex roots of $T^2 - tT + \#E(K)$. Then $\#E(L) = (\#E(K))^k + 1 - \alpha^k - \beta^k$ for all $k \geq 1$.

A polynomial time algorithm for counting the number of points in $E(L)$ was given by Schoof. The idea is to compute t modulo small primes and then use the Chinese Remainder theorem. Schoof's algorithm was improved by Elkies and Atkin and the algorithm is referred to as the *SEA* algorithm in the literature whereby $\#E(\mathbb{F}_p)$

can be computed in time $O((\log p)^6)$ by SEA algorithm. Subsequently work has been done for computing points on elliptic curves over different fields.

In the case of characteristic two curves, $K = \mathbb{F}_2$ and L is a degree k extension of K such that $\#L = 2^k$. The so-called *Koblitz curves* are given by the equation

$$E : y^2 + xy = x^3 + ax^2 + 1, \quad a \in \{0,1\}.$$

These curves were proposed by Koblitz for reasons of efficiency and for security reasons k is taken to be a prime. The following result gives the number of points in $E(L)$.

$$\#E(L) = 2^k - \left(\frac{-1+\sqrt{-7}}{2}\right)^k - \left(\frac{-1-\sqrt{-7}}{2}\right)^k + 1.$$

The *structure* of the group of points on an elliptic curve is well known. Let E be an elliptic curve defined over K. Then

$$E(K) \cong \mathbb{Z}_{n_1} \oplus \mathbb{Z}_{n_2},$$

where $n_2|n_1$ and $n_2|(\#K-1)$. As a consequence $E(K)$ is cyclic if and only if $n_2 = 1$. A point $P \in E$ is an n-torsion point if $nP = \mathscr{O}$ and $E[n]$ is the set of all n-torsion points. It is known that if $\gcd(n,q) = 1$, then $E[n] \cong Z_n \oplus Z_n$.

An elliptic curve E/K is *supersingular* if $p|t$ where $t = \#K + 1 - \#E(K)$. A result due to Waterhouse states that E/K is supersingular if and only if $t^2 = 0, \#K, 2\#K, 3\#K$ or $4\#K$.

Inversion-free arithmetic. An important issue regarding the efficient implementation of elliptic curve arithmetic is the inversion operation over the underlying finite field. According to the addition and doubling formulae described above, an inversion over the underlying field is required. As discussed earlier, an inversion can be significantly slower compared to a field multiplication. The inversion operation can be avoided by using a different coordinate system which provides a more redundant representation of a point.

The representation of P as a pair of finite field elements is said to be in affine coordinates. There are several alternatives including the so-called projective and Jacobian coordinates. In the projective coordinate system, a point is given by a triplet (X,Y,Z) which represents the affine point $(X/Z,Y/Z)$. In the Jacobian coordinate system, the triplet (X,Y,Z) represents the affine point $(X/Z^2,Y/Z^3)$. Group operations in both projective and Jacobian coordinate systems avoid inversions and the operations in the Jacobian system is faster. Below we provide the details of the group operation in the Jacobian system.

Let the curve equation be $y^2 = x^3 + ax + b$ and suppose (X_1,Y_1,Z_1) is doubled to obtain (X_3,Y_3,Z_3). Then

$$x_3 = \frac{(3X_1^2 + aZ_1^4)^2 - 8X_1Y_1^2}{4Y_1^2Z_1^2}$$

$$y_3 = \frac{3X_1^2 + aZ_1^4}{2Y_1Z_1}\left(\frac{X_1}{Z_1^2} - X_3'\right) - \frac{Y_1}{Z_1^3}$$
$$X_3 = (3X_1^2 + aZ_1^4)^2 - 8X_1Y_1^2$$
$$Y_3 = (3X_1^2 + aZ_1^4)(4X_1Y_1^2 - X_3) - 8Y_1^4$$
$$Z_3 = 2Y_1Z_1.$$

For addition, suppose (X_1, Y_1, Z_1) and $P = (X, Y, 1)$ are added to obtain (X_3, Y_3, Z_3). This is called mixed addition, since the point P is actually given by affine coordinates.

$$x_3 = \left(\frac{Y - \frac{Y_1}{Z_1^3}}{X - \frac{X_1}{Z_1^2}}\right)^2 - \frac{X_1}{Z_1^2} - X$$

$$y_3 = \left(\frac{YZ_1^3 - Y_1}{(XZ_1^2 - X_1)Z_1}\right)\left(\frac{X_1}{Z_1^2} - X_3'\right) - \frac{Y_1}{Z_1^3}$$

$$X_3 = x_3 Z_3$$
$$= (YZ_1^3 - Y_1)^2 - X_1(XZ_1^2 - X_1)^2 - X(XZ_1^2 - X_1)^2 Z_1^2$$
$$= (YZ_1^3 - Y_1)^2 - (XZ_1^2 - X_1)^2(X_1 + XZ_1^2)$$

$$Y_3 = y_3 Z_3$$
$$= (YZ_1^3 - Y_1)((XZ_1^2 - X_1)^2 X_1 - X_3) - Y_1(XZ_1^2 - X_1)^3$$
$$Z_3 = (XZ_1^2 - X_1)Z_1.$$

Scalar multiplication. The basic operation is that of scalar multiplication. Given a point P of order r and an integer $a \in \mathbb{Z}_r$, the task is to compute the a-fold of P which is written as $[a]P$ or more simply as aP. The basic left-to-right "double-and-add" algorithm is used. Recall that for left-to-right method, addition is always by P which is usually given in affine coordinates. This underlines the importance of mixed addition. There are many important issues regarding efficient and secure scalar multiplication. These are scattered throughout the literature and we refer the reader to [103] for a good idea of some of these issues.

Other forms of elliptic curves. We have mentioned only the Weierstraß form of an elliptic curve. There are other curve forms. The Montgomery form is $ay^2 = x^3 + bx + x$, $a \neq 0$ and this allows x-coordinate only scalar multiplication. The (twisted) Edwards form is $ax^2 + y^2 = 1 + dx^2y^2$; $a, d \neq 0$, $a \neq d$, which allows complete (and hence unified) formulae for addition and doubling. Among the other important forms are the Jacobi quartic form.

Finding a random point of an elliptic curve. Before we can perform cryptographic operations, we need to be able to find at least one point in the required group. The basic idea for doing this is the following: choose a random x; compute $z = x^3 + ax + b$ and find a square root y of z (if one exists). This brings us to the problem of computing square roots modulo a prime p. Checking whether an element has a square root modulo p can be done by computing the Legendre symbol $\left(\frac{z}{p}\right)$ which is 1 if z is a square mod p and -1 otherwise. The Legendre symbol is computed using the law of quadratic reciprocity. For computing square roots modulo a prime power p^e one first computes a square root modulo p and then uses "Hensel lifting" to obtain a square root modulo p^e.

Finding square roots modulo a prime is an old problem. Deterministic algorithms are known in certain cases, whereas effective randomised algorithms are known for the other cases. The basic method is due to Lagrange and works if $p \equiv 3 \bmod 4$:

$$(\pm z^{(p+1)/4})^2 = z^{(p+1)/2} = z \times z^{(p-1)/2} = z$$

since $z^{(p-1)/2} = \left(\frac{z}{p}\right) = 1$. If $p \equiv 5 \bmod 8$, then a modification of the above due to Legendre can be used. On the other hand, for $p \equiv 1 \bmod 8$ there are no known deterministic algorithms. If a quadratic nonresidue mod p is known, then a method due to Tonelli and Cipolla can be used. There are no known deterministic method for finding quadratic nonresidues. But, since about half the numbers modulo p are quadratic nonresidues, finding one can be easily done using a randomised method: choose a number α in $\mathbb{Z}_p^*$ and compute the Legendre symbol $\left(\frac{\alpha}{p}\right)$; if this is -1, then α is a quadratic nonresidue. In about two trials, we can expect to obtain such a number.

Finding a generator of a prime subgroup. Let E be an elliptic curve over a finite field K and L be a finite extension of K. Let r be a prime such that $r|\#E(L)$. Then there is a cyclic subgroup $\langle P \rangle$ of $E(L)$ which is of order r. The problem is to find a generator P of this group. The following method is used to find P. Let $r_1 = \#E(L)/r$; choose a random point R of $E(L)$; then with high probability $P = r_1 R$ is a point of order r; r being a prime, P is a generator of the required subgroup and it is possible to do cryptography over this group.

Mapping into an elliptic curve. Some cryptographic protocols require a function which maps an arbitrary string into an elliptic curve point. We briefly describe an efficient technique to hash into an element of G_1 where the order of G_1 is r (a prime). For simplicity, we consider a general characteristic field. Assume that the curve is defined over $\mathbb{Z}_p$ and is given by the short Weierstraß form $E(\mathbb{F}_p) : y^2 = x^3 + ax + b$, i.e., x and y are elements of $r\mathbb{F}_p$. Let H be a collision resistant hash function (i.e., a function for which it is computationally difficult to find distinct v_1 and v_2 such that $H(v_1) = H(v_2)$) which maps into $\mathbb{Z}_p$. Given a string str, let $x_i = H(\text{str}||i)$ for $i = 0, 1, \ldots$, and $z_i = x_i^3 + ax_i + b$. Then within a small range of values of i, it is *likely* that one of the z_i's will be a square $\mathbb{F}_p$. For the first such i, let y_i be the square root of z_i with $y_i < p/2$ (one of the square roots will be less than $p/2$ and the other will be greater). Then $P' = (x_i, y_i)$ is a point on the elliptic curve $E(\mathbb{F}_p)$. Let $\#E(\mathbb{F}_p)$ denote the number of points on $E(\mathbb{F}_p)$, then the point corresponding to the string str is obtained by multiplying P' by the cofactor $\#(E(F_p))/r$ to obtain a r-torsion point $P \in E(\mathbb{F}_p)$. This method requires computing a few Legendre symbols and computing one square root. A similar approach gives an efficient method to hash into G_2 in the case of Type 3 pairing. However, no method is known to securely hash into G_2 when we are in the Type 2 setting (see [88] for a more detailed discussion).

Note that, the approach outlined above is probabilistic in nature, though the probability of failure is extremely small. On the other hand, Icart [107] has recently suggested a method to deterministically hash into an elliptic curve.

To summarise, elliptic curves defined over finite fields provide rich examples of abelian groups. The main advantage is that no (generic) sub-exponential algorithm is known for solving the discrete log problem (we will qualify this statement later). Consequently, one can work over comparatively *smaller* fields. The gains from doing this is two-fold: the finite field arithmetic is more efficient and the storage requirement is smaller which is important for implementation on resource constrained devices.

3.1.4 Tate Pairing

We now provide a brief introduction to Tate pairing defined over elliptic curve groups. Readers are referred to the excellent exposition on pairing in [87] for more details.

Let K be a finite field and E is an elliptic curve over K. Let L be a finite degree extension of K. The group of divisors of $E(L)$ is the free abelian group generated by the points of $E(L)$. Thus any divisor D over L is of the form

$$D = \sum_{P \in E(L)} n_P(P)$$

where $n_P \in Z$ and $n_P = 0$ except for finitely many P's. The degree of a divisor is defined to be $\sum n_P$ and the divisor is said to be of degree zero if $\sum n_P = 0$.

The notion of a rational function f on an elliptic curve is a bit complicated to define. But for Weierstraß form this can be roughly understood to be the ratio of two polynomials over the curve. The divisor of a rational function f over L is defined to be

$$\mathrm{div}(f) = \sum_{P \in E(L)} \mathrm{ord}_P(f)(P)$$

where $\mathrm{ord}_P(f)$ is the order of the zero/pole that f has at P. A divisor D is said to be *principal* if $D = \mathrm{div}(f)$, for a rational function f.

An important result [163] relates principal divisors to rational functions. A divisor (over L) $D = \sum_{P \in E(L)} n_P(P)$ is principal if and only if $\sum n_P = 0$ and $\sum n_P P = \mathcal{O}$. Two divisors D_1 and D_2 are said to be *equivalent* $(D_1 \sim D_2)$ if $D_1 - D_2$ is principal. Any divisor $D = \sum n_P(P)$ of degree zero is equivalent to a (unique) divisor of the form $\langle Q \rangle - \langle \mathcal{O} \rangle$ for some $Q \in E(L)$. If $P = (x,y)$, then by $f(P)$ we mean $f(x,y)$.

If L is a degree k extension of K, then the set $\mu_r(L)$ is defined to be the cyclic subgroup of L^* of order r. Here r is prime and $r|(\#L - 1)$. The set of L-rational,

n-torsion points of E over K is defined to be the set

$$E(L)[n] = \{P \in E(L) : nP = \mathcal{O}\}.$$

Define $E(L)/rE(L)$ to be the collection of all cosets of $E(L)$ modulo $rE(L)$ and $f_{s,P}$ to be an L-rational function with divisor

$$\text{div}(f_{s,P}) = s(P) - ([s]P) - (s-1)(\mathcal{O}).$$

Let E be an elliptic curve defined over $\mathbb{F}_q$ and r be coprime to q and $r|\#E(\mathbb{F}_q)$. The embedding degree of E with respect to r is defined to be (under certain technical conditions) the smallest positive integer k such that $r|(q^k-1)$. Then k is also the least positive integer such that the field $\mathbb{F}_{q^k}$ contains all the rth roots of unity.

Let k be the embedding degree of $E/\mathbb{F}_q$ with respect to r. The (reduced and normalised) Tate pairing is defined as follows.

$$e: \ E(\mathbb{F}_q)[r] \times E(\mathbb{F}_{q^k})/rE(\mathbb{F}_{q^k}) \to \mu_r(\mathbb{F}_{q^k})$$

is given by

$$e(P,Q) = f_{r,P}(Q)^{(q^k-1)/r}, \ \text{where}$$

- P is an r-torsion point from $E(\mathbb{F}_{q^k})$;
- Q is any point in a coset in $E(\mathbb{F}_{q^k})/rE(\mathbb{F}_{q^k})$ and it can be shown that the pairing value is independent of the coset representative;
- the result is an element of $\mathbb{F}_{q^k}$ of order r.

Note that P is from $E(K)$ while Q is from $E(L)$ where K is a finite field and L is a degree k extension of L. Since P is an r-torsion point, it follows that $rP = \mathcal{O}$ and so

$$\begin{aligned}\text{div}(f_{r,P}) &= r(P) - ([r]P) - (r-1)(\mathcal{O}) \\ &= r(P) - r(\mathcal{O}).\end{aligned}$$

The computation of $f_{s,P}$ is using a double-and-add algorithm similar to that of scalar multiplication. Assume that E is given in Weierstraß form. Let P and R be points on E. We define the following rational functions and their divisors.

1. $\ell_{P,R}$ ($R \neq P$) is the line passing through P, R and $-(P+R)$.

$$\text{div}(\ell_{P,R}) = (P) + (R) + (-(P+R)) - 3(\mathcal{O}).$$

2. $\ell_{R,R}$ is the line passing through R and $-2R$.

$$\text{div}(\ell_{R,R}) = 2(R) + (-2R) - 3(\mathcal{O}).$$

3. $\ell_{R,-R}$ is the line passing through R and $-R$.

$$\text{div}(\ell_{R,-R}) = (R) + (-R) - 2(\mathcal{O}).$$

4. $h_{P,R}$ for $R \neq P$ is defined to be $h_{P,R} = \ell_{P,R}/\ell_{T,-T}$ where $T = R + P$.

$$\mathrm{div}(h_{P,R}) = (\ell_{P,R}) - (\ell_{T,-T}).$$

5. $h_{R,R}$ is defined to be $h_{R,R} = \ell_{R,R}/\ell_{T,-T}$ where $T = 2R$.

$$\mathrm{div}(h_{P,R}) = (\ell_{R,R}) - (\ell_{T,-T}).$$

Note that $\mathrm{div}(f_{1,P}) = (P) - (P) = 0$ and so $f_{1,P} = 1$. A recurrence for $f_{s,P}$ can be obtained as follows.

$$
\begin{aligned}
\mathrm{div}(f_{2m,P}) &= 2m(P) - (2mP) - (2m-1)(\mathscr{O}) \\
&= 2(m(P) - (mP) - (m-1)(\mathscr{O})) + 2(mP) - (2mP) - (\mathscr{O}) \\
&= 2\mathrm{div}(f_{m,P}) + 2(mP) + (-2mP) - 3(\mathscr{O}) \\
&\quad -((2mP) + (-2mP) - 2(\mathscr{O})) \\
&= 2\mathrm{div}(f_{m,P}) + \mathrm{div}(\ell_{mP,mP}) - \mathrm{div}(\ell_{2mP,-2mP}) \\
&= 2\mathrm{div}(f_{m,P}) + \mathrm{div}(h_{mP,mP}).
\end{aligned}
$$

$$
\begin{aligned}
\mathrm{div}(f_{2m+1,P}) &= (2m+1)(P) - ((2m+1)P) - 2m(\mathscr{O}) \\
&= 2m(P) - (2mP) - (2m-1)(\mathscr{O}) + (P) + (2mP) \\
&\quad -((2m+1)P) - (\mathscr{O}) \\
&= \mathrm{div}(f_{2m,P}) + (P) + (2mP) + (-(2m+1)P) - 3(\mathscr{O}) \\
&\quad -(((2m+1)P) + (-(2m+1)P) - 2(\mathscr{O})) \\
&= \mathrm{div}(f_{2m,P}) + \mathrm{div}(\ell_{2mP,P}) - \mathrm{div}(\ell_{(2m+1)P,-(2m+1)P}) \\
&= \mathrm{div}(f_{2m,P}) + \mathrm{div}(h_{P,2mP}).
\end{aligned}
$$

So, we have $\mathrm{div}(f_{2m,P}) = 2\mathrm{div}(f_{m,P}) + \mathrm{div}(h_{mP,mP})$ from which we get

$$f_{2m,P} = f_{m,P}^2 \times h_{mP,mP}.$$

Similarly, $\mathrm{div}(f_{2m+1,P}) = 2\mathrm{div}(f_{m,P}) + \mathrm{div}(h_{P,2mP})$ shows

$$f_{2m+1,P} = f_{2m,P} \times h_{P,2mP}.$$

Computing Tate pairing reduces to the following task. Given $P \in E(K)$ and $Q \in E(L)$ to compute $f_{r,P}(Q)$. This is done using Miller's algorithm [135, 136] in the following manner. Let $r_{t-1}r_{t-2}\ldots r_0$ be the binary expansion of r.

- Set $f \leftarrow 1$.
- Compute rP from left-to-right using "double and add".
- Let R be the input before the ith iteration.

  - $f \leftarrow f^2 \times h_{R,R}(Q); R \leftarrow 2R;$

  - if $r_{n-i} = 1$
    $f \leftarrow f \times h_{R,P}(Q);$

$$R \leftarrow R + P.$$

The above computation is the so-called Miller operation. The final R obtained after the full iteration of the loop is raised to the power $(q^k - 1)/r$ to get a unique element in $\mu_r(L)$. This is the final exponentiation part in the (reduced) Tate pairing.

The first paper [133] to introduce bilinear maps to cryptology considered a different map called the Weil pairing. Tate pairing was later introduced in [133, 84]. Several variants of Tate pairing such as ate and R-ate pairings are the currently known bilinear maps suitable for implementing pairing based cryptographic protocols (including IBE schemes). As a consequence, efficient implementation of pairing has become an active research area and there are important advances in different aspects. These include construction of elliptic curves suitable for pairing implementation [18, 1, 19], efficient algorithms for pairing [104, 128, 167] and specially efficient implementation of pairing, even on resource constrained devices such as sensor networks (see [143, 12, 58, 138, 30] for some recent results).

A related and equally important problem is the construction of elliptic curves over which pairings can be computed very fast. Such curves are called pairing friendly curves. Interested readers are referred to [83] which provides a taxonomy of pairing friendly curves.

Use of pairings in cryptanalysis. We will be interested in use of pairings to construct IBE and other cryptographic schemes. But, the first application of pairings in cryptology was essentially to cryptanalysis. Bilinear maps were suggested in [133, 84] to reduce discrete log problem over elliptic curves to that over finite fields.

As before, suppose that E is an elliptic curve over a finite field K and r is a prime divisor of $\#E(K)$; and k is the embedding degree of E with respect to r. Further, let L be the degree k extension of K. Then the bilinear map is of the form e : $E(K)[r] \times E(L)/rE(L) \to \mu_r(L)$. Write $G_1 = E(K)[r]$ and $G_2 = E(L)/rE(L)$ where G_1 and G_2 are seen as cyclic groups with generators P and Q respectively. Then $e(P, Q)$ is a generator of $\mu_r(L)$.

The discrete log problem in $G_1 = \langle P \rangle$ is that given R, it is required to find a such that $R = aP$. But, $e(aP, Q) = e(P, Q)^a$ and so the problem reduces to finding the discrete log of $h = e(R, Q)$ with respect to the base $g = e(P, Q)$. It was previously mentioned that there are no known sub-exponential algorithms for finding discrete log in an elliptic curve group whereas such algorithms are known for the finite fields. But, bilinear maps provide a method to convert the discrete log problem over elliptic curves to the discrete log problem over finite fields. This is the so-called MOV reduction named after Menezes, Okamoto and Vanstone who first observed such a relationship [133].

The question that now arises is how good is this method for finding discrete logs. If the embedding degree k is very large, then the size of L is also very large and the above method will not be effective. So, the security of the discrete log problem over elliptic curves depend on the value of the embedding degree. For supersingular curves $k \leq 6$ and so the discrete log problem over such curves are not much more difficult than over finite fields. On the other hand, for a randomly chosen ordinary elliptic curve the value of k is expected to be very large and so the above method is not useful [16].

The second argument Q of $e(P,Q)$ is an element $E(L)$. For certain types of curves (which includes the supersingular curves), it is possible to use a so-called distortion map and consider Q to be an element of $E(K)$. In this case, we have $G_1 = G_2$ giving rise to a *symmetric bilinear map*, i.e., $e(P,Q) = e(Q,P)$. This setting allows for an easy solution to the DDH problem in G_1. An instance of the DDH problem in G_1 is a tuple (P, aP, bP, cP) and the requirement is to verify whether $c = ab$ or whether c is a uniform random element from $\mathbb{Z}_p$ which is independent of a and b. This is easily done by checking whether $e(P, cP) = e(P, P)^c$ equals $e(aP, bP) = e(P, P)^{ab}$.

3.1.5 Types of Pairings

Let $e : G_1 \times G_2 \to G_T$ be a bilinear map defined over elliptic curve groups. Depending upon the structure of the group G_2 a bilinear pairing can be classified as one of the following three types [88].

Type-1. In this case, $G_2 = G_1$.
Type-2. In this case, there is an efficiently computable isomorphism ψ from G_2 to G_1.
Type-3. In this case, there are no known efficiently computable isomorphisms from G_2 to G_1.

Note that since G_1 and G_2 are both cyclic groups of the same prime order, they are certainly isomorphic. In Type-2 such an isomorphism can be easily computed whereas in Type-3 no such isomorphism is known. In both Type 2 and Type 3 it is assumed that there is no efficiently computable isomorphism from G_1 to G_2. Also note that in the case of Type 1 we have a symmetric pairing, i.e., $e(P,Q) = e(Q,P)$. In contrast, both Type 2 and Type 3 are asymmetric pairings.

Most of the pairing based protocols (including the IBE schemes) are usually first proposed in the symmetric (i.e., Type 1) setting as this allows a relatively simpler description of the protocol and its security argument. We are also going to follow the same approach in this monograph. However, it should be noted that asymmetric pairing is a better choice from the point of view of efficient implementation, specially at higher security levels. In the asymmetric setting, overall Type 3 is a better choice than Type 2 in terms of performance [88] and a recent work [59] showed that there is a *natural* transformation of any known cryptographic protocol and its security argument from Type 2 to Type 3 setting.

3.2 Hardness Assumptions

The basic hard problem in the setting of cyclic groups is the discrete log problem. The computational hardness of all other problems is contingent upon the hardness of the discrete log problem. In forming estimates of key sizes, the work in [130] considers the best known algorithm for solving discrete log problem to also be the best known algorithm for solving other problems. The discrete log problem in a cyclic group (written multiplicatively) is as follows.

Discrete Log (DL).
Instance. A cyclic group $\langle g \rangle$ of order p and an element $h \in \langle g \rangle$.
Task. Compute $a \in \mathbb{Z}_p$ such that $h = g^a$.

We next define another well-known basic computationally hard problem for a cyclic group.

Computational Diffie-Hellman (CDH).
Instance. A cyclic group $\langle g \rangle$ of order p and a tuple (g, g^a, g^b) where a and b are uniform random elements of $\mathbb{Z}_p$.
Task. Compute g^{ab}.

In other words, an algorithm (or an adversary) $\mathcal{A}$ for solving the CDH problem takes as input a tuple (g, g^a, g^b) and has to output g^{ab}. The advantage of $\mathcal{A}$ in solving the CDH problem is defined as follows.

$$\mathsf{Adv}^{\mathrm{CDH}}(\mathcal{A}) = \Pr\left[\mathcal{A}(g, g^a, g^b) \Rightarrow g^{ab}\right]$$

The CDH problem in $\langle g \rangle$ is said to be (ε, t)-hard if for any adversary $\mathcal{A}$ running in time at most t, $\mathsf{Adv}^{\mathrm{CDH}}(\mathcal{A}) \leq \varepsilon$. This problem has a decision version.

Decisional Diffie-Hellman (DDH).
Instance. A cyclic group $\langle g \rangle$ of order p and a tuple (g, g^a, g^b, g^c) for independent and uniform random a and b from $\mathbb{Z}_p$.
Task. Determine whether $c = ab$ or whether c is a uniform random element of $\mathbb{Z}_p$.

An algorithm (or an adversary) $\mathcal{A}$ for solving the DDH problem takes as input a tuple (g, g^a, g^b, g^c) and returns a bit. The event that $\mathcal{A}$ returns 1 is denoted by $\mathcal{A}(g, g^a, g^b, g^c) \Rightarrow 1$. The advantage of $\mathcal{A}$ in solving DDH is defined as follows.

$$\mathsf{Adv}^{\mathrm{DDH}}(\mathcal{A}) = \left| \Pr\left[\mathcal{A}(g, g^a, g^b, g^{ab}) \Rightarrow 1\right] - \Pr\left[\mathcal{A}(g, g^a, g^b, g^c) \Rightarrow 1\right] \right|.$$

As in the case of CDH, the DDH problem in $\langle g \rangle$ is said to be (ε, t)-hard if for any adversary $\mathcal{A}$ running in time at most t, $\mathsf{Adv}^{\mathrm{DDH}}(\mathcal{A}) \leq \varepsilon$.

We next turn our attention to the pairing setting and as noted in the previous section focus on the case of symmetric pairing only. Extensions to the asymmetric settings have been proposed in the literature and for a more detailed discussion on the computationally hard problems in the pairing setting the reader is referred to [47].

So far the general setting of a bilinear map was denoted as $e : G_1 \times G_2 \to G_T$. For the symmetric or Type 1 pairing we set $G = G_1 = G_2$ and use (p, G, G, G_T, e) to denote this setting. Let $G = \langle P \rangle$ and $G_T = \langle e(P, P) \rangle$. In the following we will assume that (P, G, G, G_t, e) along with the respective generators of G and G_T are publicly known and may not explicitly mention that while defining a problem instance.

The basic hard problem in the setting of bilinear maps is the Bilinear Diffie-Hellman problem which was first introduced by Boneh and Franklin in [39]. We define both the computational and decisional version of this problem.

Bilinear Diffie-Hellman (BDH).
Instance. A tuple (P, aP, bP, cP) where a, b and c are uniform random elements of $\mathbb{Z}_p$.
Task. Compute $e(P, P)^{abc}$.

Just like the case of CDH problem discussed above, the advantage of $\mathscr{A}$ in solving the BDH problem is defined as follows.

$$\mathrm{Adv}^{\mathrm{BDH}}(\mathscr{A}) = \Pr\left[\mathscr{A}(P, aP, bP, cP) \Rightarrow e(P, P)^{abc} \right]$$

The BDH problem in $\langle g \rangle$ is said to be (ε, t)-hard if for any adversary $\mathscr{A}$ running in time at most t, $\mathrm{Adv}^{\mathrm{BDH}}(\mathscr{A}) \leq \varepsilon$.

Decisional Bilinear Diffie-Hellman (DBDH).
Instance. (P, aP, bP, cP, Z) where $G = \langle P \rangle$, a, b, c are uniform random elements of $\mathbb{Z}_p$ and $Z \in G_T$.
Task. Determine whether $Z = e(P, P)^{abc}$ or whether Z is a random element of G_T.

An algorithm $\mathscr{A}$ for solving the DBDH problem takes a tuple (P, aP, bP, cP, Z) as input and returns a bit. The event that $\mathscr{A}$ returns 1 is denoted by $\mathscr{A}(P, aP, bP, cP, Z) \Rightarrow 1$. The advantage of $\mathscr{A}$ in solving DBDH is defined as follows.

$$\mathrm{Adv}^{\mathrm{DBDH}}(\mathscr{A})$$
$$= \left| \Pr\left[\mathscr{A}(P, aP, bP, cP, Z) \Rightarrow 1 | Z = e(P, P)^{abc} \right] \right.$$
$$\left. - \Pr\left[\mathscr{A}(P, aP, bP, cP, Z) \Rightarrow 1 | Z \text{ is random} \right] \right|.$$

The DBDH problem in (p, G, G, G_T, e) is said to be (ε, t)-hard if for any adversary $\mathscr{A}$ running in time at most t, $\mathrm{Adv}^{\mathrm{DBDH}}(\mathscr{A}) \leq \varepsilon$.

Below we consider several other problems. In each case, the advantage of an adversary can be formalised in the manner of BDH and the DBDH problems. To avoid repetition, we do not explicitly define these advantages.

As mentioned earlier if (p, G, G, G_T, e) is a Type-I pairing setting, then the DDH problem in G becomes easy. Given $(P, aP, bP, cP) \in G^4$, one simply checks whether $e(P, cP) \stackrel{?}{=} e(aP, bP)$. But, this does not imply that the CDH problem in G is easy. In fact, there is no known way to solve the CDH problem in G using the bilinear map e. Groups for which the DDH problem is easy but CDH is hard are called gap Diffie-Hellman groups and there is a corresponding gap Diffie-Hellman problem

(GDH). Informally speaking the problem states the the CDH problem is hard even if the DDH problem is easy.

The next problem was originally stated for a single cyclic group [37]. In the setting of Type 1 pairings this is the group G.

Decision Linear (DLIN).
Instance. A cyclic group $G = \langle P \rangle$ and a tuple (P, aP, bP, acP, bdP, Q).
Task. Determine whether $(c+d)P = Q$ or whether Q is a uniform random element of G which is independent of the other given elements.

The problems mentioned so far are static in the sense that the instances consist of a fixed number of elements of G. Instances of non-static problems can have a variable number of elements of G. The actual number is determined by a parameter denoted below by either h or q.

Bilinear Diffie-Hellman Exponent (BDHE).
Instance. A tuple $(P, aP, a^2P, \ldots, a^{h-1}P, a^{h+1}P, \ldots, a^{2h}P, Q)$ where $G = \langle P \rangle$, Q is a random element of G and a is a random element of $\mathbb{Z}_p$.
Task. Compute $e(P, Q)^{a^h}$.

Decisional Bilinear Diffie-Hellman Exponent (DBDHE).
Instance. A tuple $(P, aP, a^2P, \ldots, a^{h-1}P, a^{h+1}P, \ldots, a^{2h}P, Q, Z)$ with Q a random element of G, a a random element of $\mathbb{Z}_p$ and $Z \in G_T$.
Task. Determine whether $Z = e(P, Q)^{a^h}$ or whether Z is a uniform random element of G_T which is independent of the other elements.

Bilinear Diffie-Hellman inversion (BDHI).
Instance. A tuple $(P_1, \ldots, P_h)$ where $P_i = a^i P$ for some random $a \in \mathbb{Z}_p$.
Task. Compute $e(P, P)^{1/a}$.

Decisional Bilinear Diffie-Hellman inversion (DBDHI).
Instance. A tuple $(P_1, \ldots, P_h, Z)$ where $P_i = a^i P$ for some random $a \in \mathbb{Z}_p$ and $Z \in G_T$.
Task. Determine whether Z equals $e(P, P)^{1/a}$ or whether Z is a uniform random element of G_T which is independent of a.

Weak Decisional Bilinear Diffie-Hellman Inversion. There are two versions of this problem.

wDBDHI.

Instance. A tuple $(Q, P_1, \ldots, P_h, Z)$ where $P_i = a^i P$ for some random $a \in \mathbb{Z}_p$, Q is a random element of G and $Z \in G_T$.
Task. Determine whether Z equals $e(P, Q)^{1/a}$ or whether Z is a uniform random element of G_T which is independent of a.

wDBDHI*.

Instance. A tuple $(Q, P_1, \ldots, P_h, Z)$ where $P_i = a^i P$ for some random $a \in \mathbb{Z}_p$, Q is a random element of G and $Z \in G_T$.

Task. Determine whether Z equals $e(P, Q)^{a^{h+1}}$ or whether Z is a uniform random element of G_T which is independent of a.

It is possible to define the computational versions of these problems in the standard manner. The computational versions can be shown to be equivalent under a linear-time reduction and further an algorithm for either of these two computational problems gives an algorithm for the computational version of the BDHI problem with a tight reduction. See [35] for the details.

Truncated Decisional Augmented Bilinear Diffie-Hellman Exponent. In a manner similar to the BDHE, one defines the related q-ABDHE problem, where 'A' stands for 'augmented'. An instance is a tuple

$$(Q, \alpha^{q+2} P', P, \alpha P, \alpha^2 P, \ldots, \alpha^q P, \alpha^{q+2} P, \ldots, \alpha^{2q} P)$$

and the task is again to compute $e(P, Q)^{\alpha^{q+1}}$. A truncated version has instance

$$(Q, \alpha^{q+2} Q, P, \alpha P, \alpha^2 P, \ldots, \alpha^q P)$$

and the task is to compute $e(P, Q)^{\alpha^{q+1}}$ as before. Finally, Gentry [91] considers the decision version of this problem where the instance is

$$(Q, \alpha^{q+2} Q, P, \alpha P, \alpha^2 P, \ldots, \alpha^q P, Z)$$

and the task is to determine whether $Z = e(P, Q)^{\alpha^{q+1}}$ or whether Z is a uniform random element of G_T.

Pairings over composite order groups. The pairing setting is given by a tuple (p, G_1, G_2, G_T, e) where for Type-1 pairings, $G_1 = G_2$. The p here is a prime number and is the common order of the three groups G_1, G_2 and G_T. It is also possible to work in the setting where this common order is a composite number n which is a product of two (or more) "safe" primes [43], i.e., $n = p_1 p_2$. The instance contains only n and the factors of n are hidden. (For some applications, the factors may be part of the problem instance.) Composite order pairings have typically been suggested for Type-1 setting, i.e., $G_1 = G_2 = G$. The basic computationally hard problem in the setting of composite order pairings is the following. Given an element $Q \in G$ of order n and an element $R \in G$ of order p_1, it is required to determine whether another element $Z \in G$ is of order n or of order p_1.

A number of schemes have been proposed whose security relies on pairings over composite order groups. Recently, Freeman [82] showed how to convert many of these to schemes which can be based on pairing over prime order groups. Pursuing this further promises to be an interesting line of research.

Generic group model. This is an abstract model of computation over a group which was introduced in [141, 158]. In this model, the group operation is abstracted as an oracle query to a black box. The two operands are provided as the query input and the result of applying the group operation to the operands is provided as output of the query. This abstracts away the algebraic structure of the groups. More specifically, the actual method by which the result is computed is not available externally.

The generic group model is useful for providing lower bounds on the amount of effort needed to solve certain computational problems. Such lower bounds on the number of group operations needed to solve the discrete log problem were proved in [158]. Generic groups were introduced in the setting of pairings in [34], where it was used to validate an assumption called the strong Diffie-Hellman assumption. Since then, the generic group model has been considered whenever a new computational assumption has been introduced. The idea is to show that the new assumption is valid if one does not consider the internal algebraic structure of the group.

On the other hand, since the internal algebraic structure of the group is ignored, it can be argued that the generic group model is an abstraction which is far away from reality. Any reasonable algorithm for solving a computational problem in a particular group will most certainly try to exploit the inherent algebraic structure of the underlying groups. So, the assumption that the internal structure is not available to the algorithm is an impractical assumption. Consequently, one should consider lower bounds proved in the generic group model to be only a tentative indication of the computational hardness of the concerned problem. By no means can such a proof be considered to have actually established such hardness.

3.3 Conclusion

This chapter provided a brief introduction to the background algebra necessary for understanding and implementing pairing based IBE schemes. The description has been provided in an informal manner so that it is possible to quickly grasp the central ideas. There are some important practical issues such as side-channel resistance that have not been discussed. But, these issues are really outside the scope of the book and additional material can be found in the references as pointed out in the text.

Chapter 4
Boneh-Franklin IBE and its Variants

The first practical identity-based encryption scheme using bilinear pairing is attributed to Boneh and Franklin [39]. They also came up with the security definition of IBE and a reductionist proof that their IBE scheme is secure in the proposed security model assuming the hardness of the Bilinear Diffie-Hellman problem (BDH). Sakai, Ohgishi and Kashahara [151] had independently proposed an identity-based non-interactive key exchange scheme using bilinear maps. The method of private key extraction in [151] is identical to the private key extraction in the IBE scheme in [39]. The work of Boneh and Franklin caught immediate attention of the crypto community and spurred further research in this area.

4.1 Boneh-Franklin IBE

Construction of the IBE scheme in [39] proceeds in two steps. In the first step a scheme called BasicIdent is developed and shown to be secure in the sense of IND-ID-CPA. The security analysis of this scheme showed how to simulate key extraction queries made by an adversary. In the next step, this was further developed to obtain a scheme, called FullIdent, which is secure in the sense of IND-ID-CCA. In both schemes, certain hash functions are used and the security reduction models these hash functions as random oracles.

We first describe BasicIdent with an intuitive explanation of its security. This is followed by a more formal argument in terms of a security reduction.

Set-Up: Let $e : G \times G \to \Gamma_T$ be a symmetric bilinear pairing and P be a generator of G. Pick a random $s \in \mathbb{Z}_p^*$ and set $P_{\text{pub}} = sP$. Choose cryptographic hash functions $H_1 : \{0,1\}^* \to G^*$, $H_2 : G_T \to \{0,1\}^n$. The master secret is s and the public parameters are $\text{PP} = \langle P, P_{\text{pub}}, H_1, H_2 \rangle$.

Key-Gen: Given an identity $\text{id} \in \{0,1\}^*$, compute $Q_{\text{id}} = H_1(\text{id})$ and set the private key to $d_{\text{id}} = sQ_{\text{id}}$.

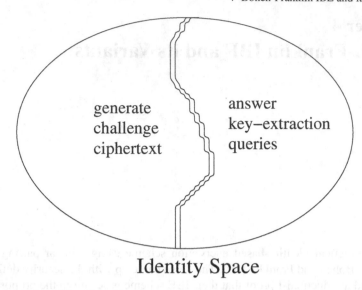

Identity Space

Fig. 4.1 Implicit partition of the identity space done by the security reduction.

Encrypt: To encrypt $M \in \{0,1\}^n$ to id compute $Q_{id} = H_1(id)$, choose a random $r \in \mathbb{Z}_p^*$ and set the ciphertext:

$$C = \langle rP, M \oplus H_2(e(Q_{id}, P_{pub})^r) \rangle$$

Decrypt: To decrypt $C = \langle U, V \rangle$ using d_{id} compute

$$V \oplus H_2(e(d_{id}, U)) = M$$

If C is an encryption of M under the public key id then we have

$$e(d_{id}, U) = e(sQ_{id}, rP) = e(Q_{id}, sP)^r = e(Q_{id}, P_{pub})^r.$$

Hence the decryption algorithm returns M given a valid encryption under the public key of id.

Security analysis of BasicIdent shows that an adversary which can win the IND-ID-CPA security game for the scheme can be used to construct an algorithm to solve an instance of the BDH problem. At a high level the idea of the proof is the following. Given an instance of the BDH problem, the challenger sets up the IBE scheme and provides the public parameters of the PKG to the adversary. The solution to the BDH problem corresponds in some sense to the master secret key of the PKG. So, the challenger does not actually know the master secret key. But, the challenger has to answer key extraction queries made by the adversary and also generate a proper challenge ciphertext. The technical difficulty of the proof is in carrying out these two tasks. This is handled by randomly partitioning the identity space into two disjoint subsets. For one part, the challenger is able to generate challenge ciphertext while for the other part the challenger is able to answer key extraction queries even without the knowledge of the master secret key of the PKG. The partitioning of the identity space is pictorially depicted in Figure 4.1. Several later works follow this basic technique of partitioning the identity space into two disjoint parts.

IND-ID-CPA security of BasicIdent is proved in [39] assuming H_1 and H_2 to be random oracles. The proof is a reduction and uses a public key encryption scheme, called BasicPub, which is defined as follows.

Key-Gen: Let P be a generator of G. Choose a random $s \in \mathbb{Z}_p$ and compute $P_{\text{pub}} = sP$; also choose a random $Q_{\text{id}} \in G^*$. Next choose a cryptographic hash function $H_2 : G_T \to \{0,1\}^n$. The private key is $d_{\text{id}} = sQ_{\text{id}}$ and the public key is $pk = \langle P, P_{\text{pub}}, Q_{\text{id}}, H_2 \rangle$.

Encrypt: Encrypt $M \in \{0,1\}^n$ as $C = \langle rP, M \oplus H_2(e(Q_{\text{id}}, P_{\text{pub}})^r) \rangle$ where r is a random element of $\mathbb{Z}_p^*$.

Decrypt: Decrypt $C = \langle U, V \rangle$ using the private key d_{id} as $V \oplus H_2(e(d_{\text{id}}, U)) = M$.

Suppose, for some identity id in BasicIdent, $H_1(\text{id})$ is mapped to Q_{id} of BasicPub. Then the Key-Gen, Encrypt and Decrypt algorithms of BasicPub essentially corresponds to the respective algorithms of BasicIdent for the identity id.

Let $\mathscr{A}_1$ be an IND-ID-CPA adversary against BasicIdent and $\mathscr{A}_2$ is an IND-CPA adversary against BasicPub, while $\mathscr{B}$ is an algorithm that solves the given BDH problem. The reduction proceeds in two steps. In the first step which we denote as Game 1, $\mathscr{A}_1$ is used to construct $\mathscr{A}_2$. In the next step which we call Game 2, $\mathscr{A}_2$ is used to construct $\mathscr{B}$. This establishes that if there is an adversary $\mathscr{A}_1$ against BasicIdent with non-negligible advantage, then one can solve the BDH problem with non-negligible probability of success. However, the existence of such a $\mathscr{B}$ contradicts the assumption that the BDH problem is computationally hard. So there is no such $\mathscr{B}$ and hence $\mathscr{A}_2$ and $\mathscr{A}_1$.

$\mathscr{B}$ plays the role of challenger in the IND-CPA game with $\mathscr{A}_2$. It runs the Key-Gen algorithm of BasicPub and gives $pk = \langle P, P_{\text{pub}}, Q_{\text{id}}, H_2 \rangle$ to $\mathscr{A}_2$. The secret key $d_{\text{id}} = sQ_{\text{id}}$ is not revealed. From pk, $\mathscr{A}_2$ passes on P, P_{pub} and H_2 to $\mathscr{A}_1$. $\mathscr{A}_2$ keeps Q_{id} to itself and uses it to simulate the random oracle H_1. The crux of the proof in the first step is in the construction of H_1.

To pose as a proper challenger to $\mathscr{A}_1$, $\mathscr{A}_2$ should be able to answer the key extraction queries and also to generate a valid challenge. While simulating H_1, $\mathscr{A}_2$ randomly partitions the identity space $\mathscr{I}$ into two disjoint subsets $\mathscr{I}_1$ and $\mathscr{I}_2$ in such a way that it is able to form a proper private key if and only if the queried identity is from $\mathscr{I}_1$. Similarly, it can form a proper challenge ciphertext if and only if the challenge identity is from $\mathscr{I}_2$. It aborts the game if the key extraction query is for an identity from $\mathscr{I}_2$ or the challenge identity is from $\mathscr{I}_1$. This in turn results in a degradation in the security reduction.

This is an intuitive explanation of the principal strategy in the security reduction of Game 1. Given this intuitive understanding, we now proceed for a more formal description. The system is setup as given in the intuitive description above. The public parameter of the PKG consists of P and P_{pub} and any query to H_2 is passed on to $\mathscr{B}$ while $\mathscr{A}_2$ simulates H_1 as follows.

Game 1

H_1-**queries:** $\mathscr{A}_1$ can query the random oracle H_1 at any time during the game. $\mathscr{A}_2$ maintains a list called H_1^{list} to answer such queries. The ith entry to the list is a 4-tuple, $\langle \text{id}_i, Q_i, b_i, c_i \rangle \in \{0,1\}^* \times G^* \times \mathbb{Z}_p^* \times \{0,1\}$. Suppose $\mathscr{A}_1$ places a query to H_1 for the identity id_j. $\mathscr{A}_2$ responds to this query in the following way.

- If id_j already exists in H_1^{list} as $\langle \text{id}_j, Q_j, b_j, c_j \rangle$ then $\mathscr{A}_2$ returns $H_1(\text{id}_j) = Q_j$.
- Otherwise $\mathscr{A}_2$ takes the following steps.

  - Generate a random $c \in \{0,1\}$ where $\Pr[c = 0] = \delta$ for some δ which is fixed a-priori for all queries.
  - Pick a random $b \in \mathbb{Z}_p^*$ and set $Q_j = bP$ if $c = 0$; otherwise set $Q_j = bQ_{\text{id}}$.
  - Add $\langle \text{id}_j, Q_j, b, c \rangle$ to H_1^{list} and return $H_1(\text{id}_j) = Q_j$ to $\mathscr{A}_1$.

Note that, based on $\Pr[c = 0] = \delta$, the identity space $\mathscr{I}$ is partitioned into two disjoint subsets $\mathscr{I}_1$ and $\mathscr{I}_2$. For identities in $\mathscr{I}_1$, we have $c = 0$ and $\mathscr{A}_2$ can answer only the private key extraction queries as we show in the next phase. While for identities in $\mathscr{I}_2$, $c = 1$ and $\mathscr{A}_2$ can generate a proper challenge ciphertext but not a private key.

Note that, whenever $\mathscr{A}_1$ places any query on H_2, $\mathscr{A}_2$ passes the query to its challenger $\mathscr{B}$. $\mathscr{A}_2$ returns whatever $\mathscr{B}$ returns.

Phase 1: Suppose $\mathscr{A}_1$ asks for the private key of id_i in the ith query. $\mathscr{A}_2$ runs the algorithm to answer the H_1-queries. Suppose $\langle \text{id}_i, Q_i, b_i, c_i \rangle$ be the corresponding tuple in H_1^{list}. If $c_i = 1$, $\mathscr{A}_2$ aborts the game. Otherwise, we have $c_i = 0$ and so $Q_i = b_iP$. Define $d_{\text{id}_i} = b_iP_{\text{pub}}$ and return d_{id_i} to $\mathscr{A}_1$. Note that, $d_{\text{id}_i} = b_isP = sQ_i$, where $H_1(\text{id}_i) = Q_i$. So this is a proper private key for id_i.

Challenge: When $\mathscr{A}_1$ decides that Phase 1 is over; it outputs a challenge identity id^* and two equal length messages M_0, M_1. id^* should not be an identity for which $\mathscr{A}_1$ placed a private key extraction query in Phase 1. $\mathscr{A}_2$ relays M_0, M_1 as its own challenge to $\mathscr{B}$. $\mathscr{B}$ chooses γ uniformly at random from $\{0,1\}$ and responds with a BasicPub ciphertext $C = \langle U, V \rangle$ of M_γ. Next, $\mathscr{A}_2$ runs the H_1-queries to find a tuple $\langle \text{id}^*, Q, b, c \rangle$ in the H_1^{list}. If $c = 0$, $\mathscr{A}_2$ aborts the game. Otherwise, $c = 1$, so $Q = bQ_{\text{id}}$ and $H_1(\text{id}^*) = Q$. $\mathscr{A}_2$ now sets $C^* = \langle b^{-1}U, V \rangle$ and returns C^* to $\mathscr{A}_1$ as the challenge ciphertext.

Let $U = rP$ and $V = M_\gamma \oplus H_2(e(Q_{\text{id}}, P_{\text{pub}})^r$ for some random $r \in \mathbb{Z}_p$. $\mathscr{A}_2$ sets $U' = b^{-1}U = b^{-1}rP = r'P$. So $e(Q_{\text{id}}, P_{\text{pub}})^r = e(b^{-1}Q, P_{\text{pub}})^r = e(Q, P_{\text{pub}})^{b^{-1}r} = e(Q, P_{\text{pub}})^{r'}$. Hence, $C^* = \langle U', V \rangle$ is a proper encryption of M_γ under the identity id^*.

Phase 2: $\mathscr{A}_1$ places additional private key extraction queries with the restriction that it cannot ask for the private key of id^*. $\mathscr{A}_2$ responds as in Phase 1.

Guess: Finally $\mathscr{A}_1$ outputs its guess γ' for γ. $\mathscr{A}_2$ relays this γ' as its own guess for γ.

If $\mathscr{A}_2$ does not abort the above game, then from the view point of $\mathscr{A}_1$ the situation is identical to that of a real attack. So we have

$$|\Pr[\gamma = \gamma'] - 1/2| \geq \varepsilon$$

where ε is the advantage of $\mathscr{A}_1$ against BasicIdent. It is easy to see that the probability that $\mathscr{A}_2$ does not abort is $\delta^q(1-\delta)$, where q is the number of key extraction queries and we want to maximize this value. This is obtained by setting $\delta = 1 - 1/(q+1)$ which in turn implies that $\Pr[\overline{\text{abort}}] \geq 1/(e \times (1+q))$, where e is the base of natural logarithms (and should not be confused with the notation $e()$ of bilinear map). Hence, $\mathscr{A}_2$'s advantage against BasicPub is at least $\varepsilon/(e \times (1+q))$. The analysis is very similar to Coron's analysis of Full Domain Hash Signature [71] and later this type of IBE has been categorized as full domain hash family.

The next step is to construct an algorithm $\mathscr{B}$ which solves the BDH problem given the adversary $\mathscr{A}_2$ against BasicPub; relating the advantage of $\mathscr{B}$ with that of $\mathscr{A}_2$. The essential intuition is, if $\mathscr{A}_2$ succeeds in the security game with a non-negligible advantage, then it must have queried the random oracle H_2 with the solution of the BDH problem as input with a non-negligible probability. This is achieved through Game 2 as described below.

Game 2 Suppose $\mathscr{B}$ is given a BDH problem instance $\langle P, aP, bP, cP \rangle$. Its task is to compute $Z = e(P,P)^{abc}$. $\mathscr{B}$ tries to solve this problem by interacting with $\mathscr{A}_2$ in the IND-CPA game.

Set-Up: $\mathscr{B}$ forms the BasicPub public key $K_{\text{pub}} = \langle P, P_{\text{pub}}, Q_{\text{id}}, H_2 \rangle$ where $P_{\text{pub}} = aP$, $Q_{\text{id}} = bP$ and H_2 is a random oracle controlled by $\mathscr{B}$. The private key $d_{\text{id}} = abP$ is unknown to $\mathscr{B}$.

H_2-queries: To respond to $\mathscr{A}_2$'s queries to the random oracle H_2, $\mathscr{B}$ maintains a list called H_2^{list}. The ith entry to the list is a tuple $\langle X_i, H_i \rangle \in G_T^* \times \{0,1\}^n$. $\mathscr{B}$ responds to any query for X_j in the following manner:

- If there is already a tuple $\langle X_j, H_j \rangle$ in H_2^{list}, then return $H_2(X_j) = H_j$.
- Otherwise, pick a random $H_j \in \{0,1\}^n$, add $\langle X_j, H_j \rangle$ to H_2^{list} and then return $H_2(X_j) = H_j$.

Challenge: $\mathscr{A}_2$ outputs two n-bit messages M_0, M_1. $\mathscr{B}$ picks a random string $R \in \{0,1\}^n$ and forms the ciphertext $C = \langle cP, R \rangle$ and returns it to $\mathscr{A}_2$. Note that, decryption of C is $R \oplus H_2(e(cP, d_{\text{id}})) = R \oplus H_2(e(P,P)^{abc})$.

Guess: $\mathscr{A}_2$ outputs its guess $\gamma' \in \{0,1\}$.
$\mathscr{B}$ picks a random tuple $\langle X_i, H_i \rangle$ from H_2^{list} and outputs X_i as the solution of the given BDH problem.

This completes the description of Game 2. What remains is to estimate the advantage of $\mathscr{B}$ against the BDH problem. Note that, in Game 2 $\mathscr{B}$ simulates a proper attack environment for $\mathscr{A}_2$. So we can estimate $\mathscr{B}$'s probability of solving the BDH problem by comparing $\mathscr{A}_2$'s behavior in the above game with its behavior in a real IND-CPA attack.

Let $\mathscr{H}$ be the event that $\mathscr{A}_2$ places a query for $Z = e(P,P)^{abc}$ to H_2 at some point in the above game. Since H_2 is modeled as a random function, without making such a query $\mathscr{A}_2$ cannot win the game with probability more than $1/2$. We state the following two facts on the event $\mathscr{H}$ (see [39] for the proofs).

Fact 1: $\Pr[\mathscr{H}]$ in the simulation in Game 2 is same as $\Pr[\mathscr{H}]$ in a real attack.

Fact 2: In a real attack $\Pr[\mathscr{H}] \geq 2\varepsilon'$ where ε' is the advantage of $\mathscr{A}_2$ against BasicPub.

In Game 2, $\mathscr{A}_2$ makes at most q_{H_2} queries to H_2. So, the probability that $\mathscr{B}$ outputs Z is at least $2\varepsilon'/q_{H_2}$.

Combining the analysis of Game 1 and 2, one comes to the final conclusion:

$$\mathsf{Adv}_{\mathscr{A}_1}^{\mathsf{BasicIdent}} \leq \frac{1}{2} e \times (1+q) q_{H_2} \varepsilon_{\mathsf{BDH}}$$

Note that, the security degrades by roughly a factor of $q q_{H_2}$ which is the product degradation in the reductions of Game 1and Game 2.

4.1.1 Security Against Chosen Ciphertext Attacks

The protocol BasicIdent, though secure against a chosen plaintext attack, is not secure against an adversary who can mount a chosen ciphertext attack. But it is possible to augment the protocol to achieve CCA security. The resulting protocol, called FullIdent, is described next.

Set-Up Define two additional hash functions $H_3 : \{0,1\}^n \times \{0,1\}^n \to \mathbb{Z}_p^*$ and $H_4 : \{0,1\}^n \to \{0,1\}^n$. Other parts are same as the Set-Up of BasicIdent.

Key-Gen: Same as the Key-Gen of BasicIdent.

Encrypt: To encrypt M to the public key id, compute $Q_{\mathsf{id}} = H_1(\mathsf{id})$, choose a random $\sigma \in \{0,1\}^n$ and set $r = H_3(\sigma,M)$. finally, the ciphertext is

$$C = \langle rP, \sigma \oplus H_2(e(Q_{\mathsf{id}}, P_{\mathsf{pub}})^r), M \oplus H_4(\sigma) \rangle$$

Decrypt: To decrypt $C = \langle U, V, W \rangle$ using the private key d_{id} do the following: compute $V \oplus H_2(e(d_{\mathsf{id}}, U)) = \sigma$; and then $W \oplus H_4(\sigma) = M$; set $r = H_3(\sigma, M)$ and verify whether $U = rP$; if not, reject C; otherwise output M.

FullIdent is the result of applying the so-called Fujisaki-Okamoto transformation [86] to BasicIdent. Let $\mathscr{E}_{pk}(M;r)$ be the encryption of M under the public key pk using random bits r where $\mathscr{E}$ is some public key encryption scheme. Fujisaki-Okamoto defined a hybrid scheme:

$$\mathscr{E}_{pk}^{hy}(M) = \langle \mathscr{E}_{pk}(\sigma, H_3(\sigma, M)), H_4(\sigma) \oplus M \rangle$$

One implication of the Fujisaki-Okamoto transformation is that if $\mathscr{E}$ is secure against chosen plaintext attack, then $\mathscr{E}^{hy}$ is secure against chosen ciphertext attack assuming that H_3 and H_4 are random oracles.

So, in a sense, Boneh-Franklin extended this idea to the identity-based setting. They apply the Fujisaki-Okamoto transformation on the IND-ID-CPA secure scheme BasicIdent to obtain the IND-ID-CCA secure scheme FullIdent.

The security reduction proceeds through several stages. First, apply the Fujisaki-Okamoto transformation on BasicPub to obtain a public key encryption scheme BasicPub^{hy}. Now, an IND-ID-CCA adversary $\mathscr{A}_1'$ against FullIdent is used to construct an IND-CCA adversary $\mathscr{A}_2'$ against BasicPub^{hy}. By direct application of the result of Fujisaki-Okamoto, there is an IND-CPA adversary $\mathscr{A}_2$ against BasicPub. Finally, by Game 2 we know that this adversary can be used to construct an algorithm to solve the BDH problem.

We give a formal description of the first reduction, i.e., given an IND-ID-CCA adversary $\mathscr{A}_1'$ against FullIdent how an IND-CCA adversary $\mathscr{A}_2'$ can be constructed against BasicPub^{hy}. Galindo in [89] showed that there is a flaw in the security reduction as described by Boneh-Franklin and suggested a remedy. The problem creeps in because of ciphertext integrity check in the Fujisaki-Okamoto transformation. We introduce the flaw in the appropriate place and then follow Galindo's modified analysis in our description.

In the IND-CCA game against BasicPub^{hy}, $\mathscr{A}_2'$ receives from its challenger the public key

$$K_{\text{pub}} = \langle P, P_{\text{pub}}, Q_{\text{id}}, H_2, H_3, H_4 \rangle$$

$\mathscr{A}_2'$ now simulates the IND-ID-CCA game for $\mathscr{A}_1'$ and interacts with its own challenger whenever necessary.

Set-Up: $\mathscr{A}_2'$ gives $\mathscr{A}_1'$ the public parameters

$$\text{PP} = \langle P, P_{\text{pub}}, H_1, H_2, H_3, H_4 \rangle$$

where H_1 is simulated in the same way as in Game 1 earlier.

Phase 1 – key extraction queries: as in game 1.

Phase 1 – decryption queries: Let $\langle \text{id}_i, C_i \rangle$ be the ith decryption query, where $C_i = \langle U_i, V_i, W_i \rangle$. If $\mathscr{A}_2'$ can generate the decryption key d_{id_i} for id_i as in the key extraction queries, then use this d_{id_i} to obtain a decryption of C_i and return the result to $\mathscr{A}_1'$. If $\mathscr{A}_2'$ is unable to form this d_{id_i} that means $c_i = 1$ in the tuple $\langle \text{id}_i, Q_i, b_i, c_i \rangle$ of H_1^{list} and hence $Q_i = b_i Q_{\text{id}}$. $\mathscr{A}_2'$ sets $C_i' = \langle b_i U_i, V_i, W_i \rangle$ and sends this to its challenger for decryption. $\mathscr{A}_2'$ returns to $\mathscr{A}_1'$ whatever the challenger returns. This is assumed to be a valid answer to the decryption query $\langle \text{id}_i, C_i \rangle$. Suppose $d_{\text{id}_i} = sQ_i$ be the private key corresponding to id_i, which is unknown to $\mathscr{A}_2'$. Let, $d_{\text{id}} = sQ_{\text{id}}$ be the decryption key of BasicPub^{hy}, which is available to the challenger. Then the FullIdent decryption of C_i using d_{id_i} would be same as the BasicPub^{hy} decryption of C_i' using d_{id}; since

$$e(b_i U_i, d_{\text{id}}) = e(b_i U_i, sQ_{\text{id}}) = e(U_i, sb_i Q_{\text{id}}) = e(U_i, d_{\text{id}_i}).$$

Galindo's Observation At this point, Galindo [89] pointed at a flaw in the original analysis of Boneh-Franklin. Galindo observed that the original argument does not take into account the fact that the decryption algorithm performs a ciphertext integrity check before returning the message. He showed that given the ciphertext $C_i' = \langle b_i U_i, V_i, W_i \rangle$, the BasicPub^{hy} decryption algorithm will reject it with overwhelming probability. Let $U_i = r_i P$ where $r_i = H_3(\sigma, M_i) \in \mathbb{Z}_p^*$, so $b_i U_i = b_i r_i P$. Since b_i is chosen at random from $\mathbb{Z}_p^*$, so $b_i r_i$ is uniformly distributed in $\mathbb{Z}_p^*$. Therefore, $H_3(\sigma, M_i) \neq b_i r_i$ with probability $(1 - 1/p)$ because H_3 is a random oracle beyond the control of $\mathscr{A}_2'$. We get a proper decryption only if $b_i = 1$, which implies $H_1(\mathrm{id}_i) = Q_{\mathrm{id}}$, where Q_{id} is part of the public key of BasicPub^{hy}.

Fixing the flaw A decryption query $\langle \mathrm{id}_i, C_i \rangle$ can be answered only if it is possible to form a proper private key corresponding to id_i or $H_1(\mathrm{id}_i) = Q_{\mathrm{id}}$. Based on this observation, Galindo suggests the following modification in the simulation.

H_1-queries: $\mathscr{A}_2'$ selects a j at random from $\{1, \ldots, q_{H_1}\}$ before initializing the H_1^{list}. When $\mathscr{A}_1'$ makes a query to H_1 at id_i with $i \neq j$, it picks a random $b_i \in \mathbb{Z}_p^*$, sets $Q_i = b_i P$, adds $\langle \mathrm{id}_i, Q_i, b_i \rangle$ to H_1^{list} and returns Q_j to $\mathscr{A}_1'$. If $i = j$, it sets $H_1(\mathrm{id}_j) = Q_{\mathrm{id}}$, adds $\langle \mathrm{id}_j, Q_{\mathrm{id}}, * \rangle$ to H_1^{list} and returns Q_{id} to $\mathscr{A}_1'$.

In terms of Figure 4.1 the identity space is partitioned in such a way that the challenge identity space now contains a single identity: id_j.

Phase 1 – Key extraction: When $\mathscr{A}_1'$ places a private key query for id_i, $\mathscr{A}_2'$ checks whether $\mathrm{id}_i = \mathrm{id}_j$. It aborts in that eventuality. Otherwise, $\mathscr{A}_2'$ forms the private key as in the original simulation and returns it to $\mathscr{A}_1'$.

Phase 1 – Decryption queries: For a decryption query $\langle \mathrm{id}_i, C_i \rangle$, if $\mathrm{id}_i = \mathrm{id}_j$, then $\mathscr{A}_2'$ asks its challenger for a decryption of C_i which it relays to $\mathscr{A}_1'$ ($\mathscr{A}_2'$ can do so because $H_1(\mathrm{id}_i) = Q_{\mathrm{id}}$). Otherwise, it forms the private key for d_{id_i} and uses it to decrypt C_i.

Challenge: At this stage $\mathscr{A}_1'$ outputs two messages M_0, M_1 and an identity id^*. If $\mathrm{id}^* \neq \mathrm{id}_j$, $\mathscr{A}_2'$ aborts the game. Otherwise, it sends M_0, M_1 to its challenger and in return gets C^* as the challenge ciphertext. C^* is an encryption of M_γ, where $\gamma \in \{0,1\}$ is chosen uniformly at random by the challenger. It returns C^* as its own challenge to $\mathscr{A}_1'$. Since, $H_1(\mathrm{id}^*) = Q_{\mathrm{id}}$, C^* is also an encryption of M_γ under id^* in FullIdent.

Phase 2: As in Phase 1, except that any key extraction query for id^*, or decryption query for $\langle \mathrm{id}^*, C^* \rangle$ is disallowed.

Guess: $\mathscr{A}_1'$ outputs its guess γ' of γ, which $\mathscr{A}_2'$ echoes.

If $\mathscr{A}_2'$ does not abort the game, then the view with respect to $\mathscr{A}_1'$ is same as that in a real IND-ID-CCA attack. The probability that $\mathscr{A}_2$ does not abort, $\Pr[\overline{abort}]$ is at least $1/q_{H_1}(1 - q_E/q_{H_1})$. Hence, $\mathscr{A}_2$'s advantage against BasicPub^{hy} is at least $\varepsilon/q_{H_1}(1 - q_E/q_{H_1}) \approx \varepsilon/q_{H_1}$.

In the next stage of the reduction, $\mathscr{A}_2'$ is used to construct an IND-CPA adversary $\mathscr{A}_2$ against BasicPub by a straight forward application of the technique of Fujisaki and Okamoto [86]. Given $\mathscr{A}_2$, the Game 2 of the previous section is used to construct an algorithm $\mathscr{B}$ that solves the BDH problem. Through this three stage reduction, an advantage ε against FullIdent can be converted to (roughly) an advantage ε/q_H^3 against BDH, where q_H is the maximum number of oracle queries to any of the random oracles H_i, $1 \le i \le 4$. In the process of fixing the flaw in security analysis, the security degradation increases because in the original analysis of Boneh-Franklin it was a factor of $1/(q_H^2 q_D)$ where q_D is the number of decryption queries and in general $q_D \ll q_H$. Galindo improved the tightness by a factor of q_H by using a second general transformation also due to Fujisaki-Okamoto [86]. Improvement by another q_H factor is obtained by using the stronger decisional version of BDH, i.e., the DBDH assumption. Later in this chapter we will describe variants of the Boneh-Franklin IBE where a simple tweak makes it possible to obtain a tight reduction.

4.2 Hierarchical Identity-Based Encryption

Soon after the appearance of the Boneh-Franklin scheme, the concept of IBE has been extended to a hierarchy of identities. Horwitz and Lynn [106] proposed a definition of HIBE and constructed two-level HIBE that achieves a limited kind of security. Gentry and Silverberg extended the BF-IBE scheme to a HIBE scheme in [94]. This scheme has been shown to be secure against adaptive identity and chosen ciphertext attack in the random oracle model. Here we describe the GS-HIBE followed by an idea of the security reduction.

BasicHIBE

Set-Up: Let P be an arbitrary generator of G. Pick a random $x_0 \in \mathbb{Z}_p^*$ and set $P_{\text{pub}} = x_0 P$. Also choose cryptographic hash functions $H_1 : \{0,1\}^* \to G$ and $H_2 : G_T \to \{0,1\}^n$. The public parameters are $\mathsf{PP} = \langle P, P_{pub}, H_1, H_2 \rangle$, while the master secret is x_0.

Key-Gen: An identity id at the jth level is represented as $\mathsf{id} = (\mathsf{id}_1, \ldots, \mathsf{id}_j)$. The PKG chooses random elements $x_1, \ldots, x_{j-1} \in \mathbb{Z}_p^*$ and computes $d_i = x_i P$ for $1 \le i \le j-1$ and $d_j = \sum_{i=1}^{j} x_{i-1} Q_i$, where $Q_i = H_1(\mathsf{id}_1, \ldots, \mathsf{id}_i)$. It gives $d_{\mathsf{id}} = (d_1, \ldots, d_j)$ to id.

A private key for id can be generated by its parent. Let $\mathsf{id}|_{j-1} = (\mathsf{id}_1, \ldots, \mathsf{id}_{j-1})$ be the parent of id, i.e., $\mathsf{id}|_{j-1}$ is one level up in the hierarchy with respect to id. Let the private key of $\mathsf{id}|_{j-1}$ be $d_{\mathsf{id}|_{j-1}} = (d_1', \ldots, d_{j-1}')$. Then d_{id} can be formed by $\mathsf{id}|_{j-1}$ as follows: compute $Q_{\mathsf{id}} = H_1(\mathsf{id}_1, \ldots, \mathsf{id}_j)$, choose a random $x_{j-1} \in \mathbb{Z}_p^*$ and set $d_j = d_{j-1}' + x_{j-1} Q_{\mathsf{id}}$ and set $d_i = d_i'$ for $1 \le i \le j-2$ and $d_{j-1} = x_{j-1} P$. The private key, $d_{\mathsf{id}} = (d_1, \ldots, d_j)$ is given to id.

Note that by iterating the above process any ancestor of id can generate a decryption key for id. In more details, the ancestor with identity $(\mathsf{id}_1, \ldots, \mathsf{id}_i)$ for $i < j$, first generates a key for the identity $(id_1, \ldots, id_i, id_{i+1})$ at the $(i+1)$st level, then a key for the identity at the $(i+2)$nd level and so on.

Encrypt: Encrypt of M to the identity $id = (id_1, \ldots, id_j)$ is done as follows: compute $Q_i = H_1(id_1, \ldots, id_i)$ for $1 \le i \le j$. Then choose a random $r \in \mathbb{Z}_p^*$ and set the ciphertext

$$C = \langle rP, rQ_2, \ldots, rQ_j, M \oplus H_2(e(P_{\mathsf{pub}}, Q_1)^r) \rangle$$

Decrypt: Given $C = \langle U_0, U_2, \ldots, U_j, V \rangle$ and $d_{id} = \langle d_1, \ldots, d_j \rangle$, compute

$$V \oplus H_2 \left(\frac{e(U_0, d_j)}{\prod_{i=2}^{j} e(d_{i-1}, U_i)} \right) = M$$

The correctness of decryption can be seen from the following computation.

$$
\begin{aligned}
\frac{e(U_0, d_j)}{\prod_{i=2}^{j} e(d_{i-1}, U_i)} &= \frac{e\left(rP, \sum_{i=1}^{j} x_{i-1} Q_i\right)}{\prod_{i=2}^{j} e(x_{i-1} P, rQ_i)} \\
&= \frac{e(rP, x_0 Q_1) \times e\left(rP, \sum_{i=2}^{j} x_{i-1} Q_i\right)}{\prod_{i=2}^{j} e(x_{i-1} P, rQ_i)} \\
&= e(P_{\mathsf{pub}}, Q_1)^r \times \frac{\prod_{i=2}^{j} e(x_{i-1} P, rQ_i)}{\prod_{i=2}^{j} e(x_{i-1} P, rQ_i)} \\
&= e(P_{\mathsf{pub}}, Q_1)^r.
\end{aligned}
$$

If the HIBE is restricted to the first level only, then this is exactly the BasicIdent scheme of Boneh and Franklin [39]. On the other hand, the above HIBE supports an arbitrary number of levels, i.e., during Set-Up, the PKG does not have to fix a priori the maximum number of levels that the system can support.

Security of BasicHIBE against chosen plaintext attack (i.e., IND-ID-CPA security) can be proved in the same manner as that of the BasicIdent of previous section. In the first stage, an IND-ID-CPA adversary, $\mathscr{A}_1$ against BasicHIBE is used to construct an IND-CPA adversary $\mathscr{A}_2$ against BasicPub (the same public key encryption scheme defined in the context of Boneh-Franklin IBE). In the next stage, this $\mathscr{A}_2$ is utilised to construct an algorithm $\mathscr{B}$ that solves the BDH problem.

The security reduction is in the adaptive setting. However, this suffers from a large degradation factor. Let's briefly see why this is so. Here, instead of a single identity, we have an identity tuple of arbitrary levels. Recall that in the reduction for BasicIdent of BF-IBE a typical entry in H_1^{list} is of the form $\langle id, Q, b, c \rangle$. In case of BasicHIBE an entry for $id = (id_1, \ldots, id_j)$ is of the form $\{\langle id_i \rangle, \langle Q_{id_i} \rangle, \langle b_i \rangle, \langle c_i \rangle\}$ where $1 \le i \le j$, i.e., each $\langle \cdots \rangle$ contains j many entries, whereas in case of BasicIdent of BF-IBE they contained only a single term.

Let the target identity be $id^* = (id_1^*, \ldots, id_h^*)$ then the corresponding entry in the H_1^{list} has terms $c_1, \ldots, c_h \in \{0, 1\}^h$. So, instead of a single c, we have to maintain h many c_is in H_1^{list}. $\mathscr{A}_2$ can generate a proper challenge ciphertext if and only if all these h c_is have the same value, 1. But these c_is are chosen independent of each other, so the probability that all are 1 is the product of the probabilities that each is 1. Hence, we get a security degradation which is exponential in the number of levels in the target identity tuple.

Once $\mathscr{A}_2$ has been constructed, the next stage of the reduction is exactly that of Game 2 in case of Boneh-Franklin IBE. Finally, in the adaptive setting one gets the following result:

$$\mathsf{Adv}_{\mathscr{A}_1}^{\mathsf{BasicHIBE}} \leq \frac{(e \times (q+h))^h q_H}{h} \varepsilon_{BDH}$$

where e is the base of natural logarithm, q (resp. q_H) is the maximum number of key extraction query (resp. random oracle query to H_2).

This means, if there is an IND-ID-CPA adversary $\mathscr{A}_1$ in the adaptive setting having advantage ε against BasicHIBE and that makes at most q_H queries to H_2 and q private key extraction queries and H_1, H_2 are random oracles then there is an algorithm $\mathscr{B}$ that solves the BDH problem with an advantage of roughly $\varepsilon/(q^h q_H)$, where h is the number of levels in the target identity. In other words, the security degrades exponentially with the number of levels of the HIBE.

Applying the Fujisaki-Okamoto transformation to the BasicHIBE, Gentry and Silverberg obtain an IND-ID-CCA secure HIBE which is called FullHIBE. The construction as well as the security proof is analogous to that of FullIdent and not detailed here. Similarly, Galindo's observation [89] with respect to FullIdent is also applicable for the scheme FullHIBE.

4.3 Boneh-Katz-Wang CPA-Secure IBE

This variant of Boneh-Franklin IBE scheme has appeared as a remark in a paper on signatures by Katz and Wang [118] and has been attributed to Boneh. The interesting feature is that with a small tweak in the construction it is possible to achieve a tight security reduction. The IBE scheme is described in Figure 4.2.

The security of this scheme can be based on the BDH problem. We provide an idea of the security reduction based on the stronger DBDH assumption. In this case, the second hash function H_2 is not required and instead $W_i = e(P_1, Q_{\mathsf{id},i})^t$, for $i \in \{0,1\}$ are directly used to mask the message. An adversary $\mathscr{A}$ for this IBE scheme can be converted to a DBDH solver $\mathscr{B}$ in the following manner. Given an instance $(P, P_1 = \alpha P, P_2 = \beta P, \gamma P, Z)$, $\mathscr{B}$ declares the public parameters to be (P, P_1). Queries to the hash oracle are tackled in the following manner. Suppose the query is (id, b). If this already occurs in the H-list (i.e., the list for the hash function H_1), then return previous value; otherwise, generate a random bit c and a random element $x \in \mathbb{Z}_p$; store (id, c, x, xP) and $(\mathsf{id}, \overline{c}, x, xP_2)$ in H-list; and return either xP or xP_2 depending on whether $b = c$ or $b = \overline{c}$. This simulation ensures that for the challenge identity id^* one of the hash values $H_1(\mathsf{id}^*, 0)$ and $H_1(\mathsf{id}^*, 1)$ will provide a proper encryption, i.e., one of C_0^* or C_1^* is proper (if $Z = e(P, P)^{\alpha\beta\gamma}$) but which one that is hidden from the adversary. So, the adversary's probability of guessing the correct message is half

Fig. 4.2 Boneh-Katz-Wang IBE Scheme.
1. Identities are arbitrary length binary strings.
2. Messages are n-bit strings.
3. KeyGen first checks whether a decryption key has been generated earlier for this identity. If it has been done, then the same key is returned.

Set-Up	KeyGen: Identity id.
1. Choose s uniformly at random from $\mathbb{Z}_p$. 2. Set $P_{pub} = sP$. 3. Let $H_1 : \{0,1\}^* \to G$ and $H_2 : G_T \to \{0,1\}^n$ be hash functions. 4. Public parameters: P, P_{pub}, H_1, H_2. 5. Master secret key: s.	1. Choose a random bit b. 2. $Q_{id} = H_1(id, b)$. 3. Output $d_{id} = (sQ_{id}, b)$.
Encrypt: Identity id; message M.	Decrypt: Identity id; ciphertext (U, C_0, C_1); decryption key $d_{id} = (V, b)$.
1. Choose t randomly from $\mathbb{Z}_p$. 2. $Q_{id,0} = H_1(id,0)$; $Q_{id,1} = H_1(id,1)$. 3. $W_0 = e(P_1, Q_{id,0})^t$; $W_1 = e(P_1, Q_{id,1})^t$; 4. Output $(tP, H_2(W_0) \oplus M, H_2(W_1) \oplus M)$.	1. $W = e(U, V)$. 2. Output $C_b \oplus H_2(W)$.

of what it would be if both C_0^* and C_1^* were proper. On the other hand, if Z is random, then both C_0^* and C_1^* are random and provide no information to the adversary about which message was chosen. These considerations show that the advantage of $\mathscr{A}$ is upper bounded by $\varepsilon_{dbdh}/2$, which constitutes a tight reduction.

4.4 Attrapadung et al's CCA-Secure IBE

This IBE scheme is a modification of the Boneh-Katz-Wang (BKW) IBE scheme to attain CCA-security. The Boneh-Katz-Wang IBE scheme is a variant of BasicIdent and hence only CPA-secure. An adversary having access to a decryption oracle can easily break the scheme. An intuitive reason for this is the following.

In the BKW scheme, there are essentially two encryptions of the message. In the CCA-security game, the challenge ciphertext contains two different encryptions of the chosen message for id$||0$ and id$||1$. Given a challenge ciphertext (U, C_0^*, C_1^*), the adversary forms $(U, C_0^* + S, C_1^* + S)$ for a random n-bit string S and queries the decryption oracle. This will return $M_\gamma + S$ where γ was the random bit chosen by the challenger. From this the adversary obtains M_γ and wins the game.

The IBE scheme by Attrapadung et al [14] attempts to achieve CCA-security by converting to hybrid encryption and providing two different encapsulations of the session key. During decryption, it is verified that these two encapsulations are of the same key. On looking at the decryption algorithm, one will notice that the value of R is recovered in two different ways and then checked for equality. Also, the randomiser t for the ciphertext is recovered and the check $U = tP$ is applied. If both these hold, then the ciphertext is well formed, i.e., the two encryptions V_0 and V_1 are of the same R and this R is used in the derivation of t. Now, the argument for CPA-security ensures that for well-formed ciphertexts, the adversary's advantage in breaking the scheme is negligible. Details of the scheme are shown in Figure 4.3.

Fig. 4.3 Attrapadung et al's IBE Scheme.
1. Identities are arbitrary length binary strings.
2. Messages are n-bit strings.
3. $\mathscr{K}$ is the key space for the symmetric encryption algorithm and E_K (resp. D_K) denotes the encryption (resp. decryption) algorithm for symmetric encryption using the key K from $\mathscr{K}$.
4. KeyGen first checks whether a decryption key has been generated earlier for this identity. If it has been done, then the same key is returned.

Set-Up	
1. Choose s uniformly at random from $\mathbb{Z}_p$. 2. Set $P_{\mathsf{pub}} = sP$. 3. Let $H : \{0,1\}^* \to G$, $\quad H_1 : \{0,1\}^* \to \{0,1\}^{k_1}$ and $\quad \widetilde{H} : \{0,1\}^* \to \mathbb{Z}_p \times \mathscr{K}$ be hash functions. 4. Public parameters: $P, P_{\mathsf{pub}}, H, H_1, \widetilde{H}$. 5. Master secret key: s.	KeyGen: Identity id. 1. Choose a random bit b. 2. $Q_{\mathsf{id}} = H(\mathsf{id}, b)$. 3. Output $d_{\mathsf{id}} = (sQ_{\mathsf{id}}, b)$.
Encrypt: Identity id; message M. 1. $Q_{\mathsf{id},0} = H(\mathsf{id},0)$; $Q_{\mathsf{id},1} = H(\mathsf{id},1)$; 2. choose R randomly from $\{0,1\}^{k_1}$; 3. $t \| K = \widetilde{H}(R, \mathsf{id})$; $U = tP$; 4. $W_0 = e(P_{\mathsf{pub}}, Q_{\mathsf{id},0})^t$; $W_1 = e(P_{\mathsf{pub}}, Q_{\mathsf{id},1})^t$; 5. $V_0 = H_1(W_0, \mathsf{id}, 0) \oplus R$; $V_1 = H_1(W_1, \mathsf{id}, 1) \oplus R$; 6. $\mathsf{cpr} = E_K(M)$; 7. Output $(U, V_0, V_1, \mathsf{cpr})$.	Decrypt: Identity id; ciphertext $(U, V_0, V_1, \mathsf{cpr})$; decryption key $d_{\mathsf{id}} = (V, b)$. 1. $W = e(U, V)$; $R = V_b \oplus H_1(W, \mathsf{id}, b)$; 2. $t \| K = \widetilde{H}(R, \mathsf{id})$; 3. $W' = e(P_{\mathsf{pub}}, Q_{\mathsf{id}, \overline{b}})^t$; 4. $R' = V_{\overline{b}} \oplus H_1(W', \mathsf{id}, \overline{b})$; 5. if $R \neq R'$ or $U \neq tP$ return $\perp$; 6. $M = D_K(\mathsf{cpr})$; 7. Output M.

4.5 Conclusion

This chapter introduces the first practical construction of IBE, namely the Boneh-Franklin IBE. We first describe the CPA-secure version and its security argument. Next comes the CCA-secure version of BF-IBE and its security argument. The concept of IBE was soon generalized to HIBE. The Gentry-Silverberg HIBE is presented along with an intuitive discussion of its security. Several researchers tried to improve upon the tightness of security reduction of BF-IBE. Two such variants are

described here. Coron [72] also proposed a different variant of BF-IBE with tight
reduction which we have not described.

Chapter 5
Selective-Identity Model

The (H)IBE schemes described in the previous section have one common feature. For the security reduction, all of them assume certain hash functions to be independent and uniform random functions, i.e., these are modelled as so-called *random oracles*. Clearly, no practical hash function can satisfy this condition. Due to this reason, the security assurance provided by proofs in the random oracle model has been questioned from a theoretical perspective. Without going into the details of this debate, the following question is still meaningful and important. Are there any cryptographic protocols whose security is based on a standard computational assumption but the security reduction does not depend on the random oracle model?

In the context of IBE, this question was first addressed by Canetti, Halevi and Katz [53]. They provided a positive answer to the question. There was, however, a trade-off. To prove security, they had to weaken the security model for (H)IBE schemes. They introduced the selective-identity security game and obtained proofs under this game. See Section 2.3.3 for the details of the selective-identity security model.

Canetti, Halevi and Katz [53] introduced the notion of binary tree encryption (BTE). This is different from HIBE. Each node in BTE has exactly two children – the left child and the right child. A public key encryption scheme is described and they also show [53] how an (H)IBE can be constructed from a BTE. This provided the first (H)IBE construction that is secure without random oracles in the selective-identity model. We do not provide the details of their construction and security proof. The construction is more of a theoretical step in removing random oracles. A practical limitation of the construction is its large computational overhead – during decryption, for each bit of the identity one pairing computation is required. In this chapter, we discuss some of the later important (H)IBE schemes that have been proved to be secure in the selective-identity model.

Boneh and Boyen [32] proposed two efficient IBE schemes that are secure in the selective-identity model without random oracle. The first construction was presented as an HIBE which we denote as BB-HIBE and the corresponding IBE will be called BB-IBE. We give a detailed description of the BB-(H)IBE protocol and its security reduction below. Though proven secure in a very restrictive security model, the importance of BB-(H)IBE lies in the fact that the algebraic techniques introduced in the construction and security reduction turned out to be extremely useful in later works.

The other construction that we describe is a HIBE scheme due to Boneh, Boyen and Goh [35]. This provided a HIBE scheme where the size of the ciphertext does not depend on the number of components in the identity tuple. Such a scheme can be called a constant size ciphertext HIBE scheme. (Note, however, that the size of the ciphertext still does depend on the targeted security level, since the security level determines the size of the underlying groups and hence the length of representation of the group elements.)

5.1 Boneh-Boyen HIBE

Here individual components of an identity tuple are considered to be elements of $\mathbb{Z}_p$. In practice, identities will be arbitrary bit-strings. A practical implementation will use a collision-resistant hash function to map bit-strings to elements of $\mathbb{Z}_p$.

Set-Up: Select a random generator $P \in G$, a random $x \in \mathbb{Z}_p$ and set $P_1 = xP$. Also pick random elements $Q_1, \ldots, Q_h, P_2 \in G$. Then

$$PP = (P, P_1, P_2, Q_1, \ldots, Q_h); \quad msk = xP_2$$

The maximum height of the HIBE is h and the construction yields an IBE when $h = 1$. Define publicly computable family of functions $F_j : \mathbb{Z}_p \to G$ for $j \in \{1, \ldots, h\}$: $F_j(\alpha) = \alpha P_1 + Q_j$.

Key-Gen: Given an identity $\mathsf{id} = (\mathsf{id}_1, \ldots, \mathsf{id}_j)$ of depth $j \leq h$, pick random $r_1, \ldots, r_j$ from $\mathbb{Z}_p$ and compute

$$d_{\mathsf{id}} = \left(xP_2 + \sum_{i=1}^{j} r_i F_i(\mathsf{id}_i), r_1 P, \ldots, r_j P \right).$$

The decryption key d_{id} consists of $(j+1)$ elements of G. For $j = 1$, i.e., when the identity tuple consists of a single component id_1, we have

$$d_{\mathsf{id}_1} = (xP_2 + r_1(\mathsf{id}_1 P_1 + Q_1), r_1 P).$$

Note that d_{id} can be generated given the private key $d_{\mathsf{id}|j-1} = (d_0, d_1, \ldots, d_{j-1})$ of $\mathsf{id}|_{j-1} = (\mathsf{id}_1, \ldots, \mathsf{id}_{j-1})$. Generate a random $r_j \in \mathbb{Z}_p$ and set

$$d_{\mathsf{id}} = (d_0 + r_j F_j(\mathsf{id}_j), d_1, \ldots, d_{j-1}, r_j P).$$

Encrypt: Encrypt $M \in G_T$ for $\mathsf{id} = (\mathsf{id}_1, \ldots, \mathsf{id}_j)$ as

$$C = (e(P_1, P_2)^s \times M, sP, sF_1(\mathsf{id}_1), \ldots, sF_j(\mathsf{id}_j))$$

where s is a random element of $\mathbb{Z}_p$. Note that the length of the ciphertext depends on the length of the identity.

Decrypt: Decrypt $C = (A, B, C_1, \ldots, C_j)$ using the private key $d_{\mathsf{id}} = (d_0, d_1, \ldots, d_j)$ as

$$A \times \frac{\prod_{i=1}^{j} e(C_i, d_i)}{e(B, d_0)}.$$

Note that for a proper ciphertext, $A = M \times e(P_1, P_2)^s$, i.e., the message M has been masked by the value $e(P_1, P_2)^s$. Since s is a random value which is independent of the other random quantities, this effectively results in A being a random value from G_T. For a properly formed ciphertext, the correctness of the decryption procedure follows from the following computation.

$$\frac{\prod_{i=1}^{j} e(C_i, d_i)}{e(B, d_0)} = \frac{\prod_{i=1}^{j} e(r_i F_i(\mathsf{id}_i), P)^s}{e\left(sP, xP_2 + \sum_{i=1}^{j} r_i F_i(\mathsf{id}_i)\right)}$$

$$= \frac{\prod_{i=1}^{j} e(F_i(\mathsf{id}_i), P)^{r_i s}}{e(P, P_2)^{sx} \times \prod_{i=1}^{j} e(P, F_i(\mathsf{id}_i)^{r_i s})}$$

$$= \frac{1}{e(P_1, P_2)^s}.$$

Also note that the blinding factor sx commutes under pairing allowing more than one ways to derive $e(P_1, P_2)^s$. This was later termed as commutative blinding and Several (H)IBE schemes utilized this framework.

5.1.1 Security

CPA security of BB-HIBE is proved in the selective-identity model. As is usual for security reductions, the idea is to use an adversary for attacking the HIBE in the selective-identity model to construct an algorithm for solving the DBDH problem. An adversary $\mathscr{A}$ for the selective-identity game has to commit to an identity tuple before the system is set-up. The essential idea is to form the public parameters using the target identity tuple and the DBDH instance in such a way that all the key extraction queries of $\mathscr{A}$ (except on the target identity or any of its prefix) can be answered by $\mathscr{A}$. A valid challenge, on the other hand, can be generated for the target identity only. So, based on the target identity, the identity space is partitioned into two disjoint subsets.

Below we provide the details of the proof. We would like to draw the reader's attention to the particular method of simulating a key extraction query made by the adversary.

Initialization: $\mathcal{A}$ commits to a target identity $\text{id}^* = (\text{id}_1^*, \ldots, \text{id}_{h'}^*)$ of height $h' \leq h$. If $h' < h$, $\mathcal{B}$ adds extra random elements from $\mathbb{Z}_p$ to make id^* an identity of height h. Let us denote these extra $(h - h')$ elements by $\text{id}_{h'+1}^*, \ldots, \text{id}_h^*$.

Set-Up: Given a DBDH instance (P, aP, bP, cP, Z), $\mathcal{B}$ sets $P_1 = aP$ and $P_2 = bP$. It then picks random $\alpha_1, \ldots, \alpha_h \in \mathbb{Z}_p$ and defines $Q_j = \alpha_j P - \text{id}_j^* P_1$ for $1 \leq j \leq h$. It gives $\mathcal{A}$ the public parameters $\text{PP} = (P, P_1, P_2, Q_1, \ldots, Q_h)$. Note that, the msk $= aP_2 = abP$ is unknown to $\mathcal{B}$. Define the function $F_j(x) = xP_1 + Q_j = (x - \text{id}_j^*)P_1 + \alpha_j P$ for $1 \leq j \leq h$.

Phase 1: $\mathcal{A}$ makes up to q private key queries. In a private key query corresponding to an identity $\text{id} = (\text{id}_1, \ldots, \text{id}_u)$, with $u \leq h$ the only restriction is that id is not a prefix of id^*. Let, j be the smallest index such that $\text{id}_j \neq \text{id}_j^*$. $\mathcal{B}$ chooses random $r_1, \ldots, r_j \in \mathbb{Z}_p$ and first computes

$$
\begin{aligned}
d_{0|j} &= \frac{-\alpha_j}{(\text{id}_j - \text{id}_j^*)} P_2 + r_j F_j(\text{id}_j) \\
&= \frac{-\alpha_j}{(\text{id}_j - \text{id}_j^*)} P_2 + r_j((\text{id}_j - \text{id}_j^*)P_1 + \alpha_j P) \\
&= abP - abP + \frac{-\alpha_j}{(\text{id}_j - \text{id}_j^*)} bP + r_j((\text{id}_j - \text{id}_j^*)P_1 + \alpha_j P) \\
&= aP_2 + \left(r_j - \frac{b}{\text{id}_j - \text{id}_j^*} \right)((\text{id}_j - \text{id}_j^*)P_1 + \alpha_j P) \\
&= aP_2 + \tilde{r}_j F_j(\text{id}_j)
\end{aligned}
$$

where $\tilde{r}_j = r_j - \frac{b}{\text{id}_j - \text{id}_j^*}$. So $\mathcal{B}$ forms the private key of $(\text{id}_1, \ldots, \text{id}_j)$ as

$$
d_0 = d_{0|j} + \sum_{i=1}^{j-1} r_i F_i(\text{id}_i), d_1 = r_1 P, \ldots, d_{j-1} = r_{j-1} P, d_j = -\frac{1}{\text{id}_j - \text{id}_j^*} P_2 + r_j P = \tilde{r}_j P
$$

It is easy to verify that $(d_0, d_1, \ldots, d_j)$ is a valid private key for $(\text{id}_0, \ldots, \text{id}_j)$. Once the private key of $(\text{id}_1, \ldots, \text{id}_j)$ is formed, $\mathcal{B}$ uses the **Key-Gen** algorithm to form a private key for id and returns it to $\mathcal{A}$.

Note that, $\mathcal{B}$ can derive a valid private key for an identity id without the knowledge of the master secret. This is possible as long as id is not a prefix of id^*. The above algebraic technique of private key derivation is one of the major technical novelties introduced by Boney and Boyen [32]. Recall that, if the original target identity $\text{id}^* = (\text{id}_1^*, \ldots, \text{id}_{h'}^*)$ is of height less than h, then $\mathcal{B}$ augments it to an h-tuple by randomly choosing $\text{id}_{h'+1}^*, \ldots, \text{id}_h^*$. This forms $h - h'$ descendants of id^* as $(\text{id}^*|\text{id}_{h'+1}^*), (\text{id}^*|\text{id}_{h'+1}^*, \text{id}_{h'+2}^*), (\text{id}^*|\text{id}_{h'+1}^*, \text{id}_{h'+2}^*, \ldots, \text{id}_h^*)$. Hence, $\mathcal{B}$ cannot generate the private key of any of these descendants of id^*. But $\mathcal{A}$ can ask for the private key of any of them. For example, $\text{id} = (\text{id}_1^*, \ldots, \text{id}_{h'}^*, \text{id}_{h'+1}^*, \ldots, \text{id}_h^*)$ can be a valid query for private key extraction. In such eventuality, $\mathcal{B}$ has to abort the game. Since

$\text{id}^*_{h'+1}, \ldots, \text{id}^*_h$ are chosen by the simulator uniformly at random, the probability of abort is very low – of the order q/p.

Challenge: After completion of Phase 1, $\mathscr{A}$ outputs two messages $M_0, M_1 \in G_T$. $\mathscr{B}$ chooses a random bit γ and forms the ciphertext $C = (M_\gamma \cdot Z, cP, \alpha_1 cP, \ldots, \alpha_{h'} cP)$. Note that, as per construction, $F_i(\text{id}^*_i) = \alpha_i P$ for $1 \leq i \leq h'$, so

$$C = (M_\gamma \cdot Z, cP, cF_1(\text{id}^*_1), \ldots, cF_{h'}(\text{id}^*_{h'})).$$

If $Z = e(P,P)^{abc} = e(P_1, P_2)^c$ then C is a valid encryption of M_γ. On the other hand, if Z is random, then C is an encryption of a random element in G_T.

Phase 2: $\mathscr{A}$ makes additional queries which $\mathscr{B}$ answers just like Phase 1. Total number of queries in Phase 1 and 2 together should not exceed q.

Guess: Eventually, $\mathscr{A}$ outputs its guess γ' of γ. If $\gamma' = \gamma$, $\mathscr{B}$ outputs 1, otherwise it outputs 0.

When $Z = e(P,P)^{abc}$, then $\mathscr{A}$'s view in the above game is identical to that in a real attack. In that case $|\Pr[\gamma = \gamma'] - 1/2| \geq \varepsilon$. On the other hand if Z is a random element of G_T then $\Pr[\gamma = \gamma'] = 1/2$. Since the events $Z = e(P,P)^{abc}$ and Z is random are equiprobable, it is easy to see that

$$\text{Adv}^{\text{DBDH}}_{\mathscr{B}} \geq \frac{\varepsilon}{2}$$

In other words, if the (t, ε)-DBDH assumption holds in G, G_T then the h-HIBE of Boneh-Boyen is $(t', q, 2\varepsilon)$-IND-sID-CPA secure for arbitrary h and q and any $t' < t - O(\tau h q)$ where τ is the time for a scalar multiplication in G.

5.2 Constant Size Ciphertext HIBE

Boneh, Boyen and Goh [35] proposed a HIBE scheme where the length of the ciphertext does not depend on the number of components in the identity tuple. We refer to this scheme as BBG-HIBE. The construction is described below.

Identities at a depth u are of the form $(\text{id}_1, \ldots, \text{id}_u) \in (\mathbb{Z}^*_p)^u$. Messages are elements of G_T. As mentioned in the context of BB-(H)IBE, if the identity components are arbitrary binary strings then a collision-resistant hash function is used to map them to elements of $\mathbb{Z}^*_p$.

Set-Up: Let $\langle P \rangle = G$. Choose a random $x \in \mathbb{Z}_p$ and set $P_1 = xP$. Choose random elements $P_2, P_3, Q_1, \ldots, Q_h \in G$.

Set the public parameter as $\text{PP} = (P, P_1, P_2, P_3, Q_1, \ldots, Q_h)$ while the master key is xP_2.

Key-Gen: Given an identity $\text{id} = (\text{id}_1, \ldots, \text{id}_k) \in (\mathbb{Z}^*_p)^k$ of depth $k \leq h$, pick a random $r \in \mathbb{Z}_p$ and output

$$d_{\mathsf{id}} = (xP_2 + r(\mathsf{id}_1 Q_1 + \cdots + \mathsf{id}_k Q_k + P_3), rP, rQ_{k+1}, \ldots, rQ_h).$$

The private key for id can also be generated given the private key of any of its ancestor $\mathsf{id}_{|j}$, $j < k$. We explain this process for the immediate ancestor $\mathsf{id}_{|k-1}$ of id as follows: Let $d_{\mathsf{id}|k-1}$ be the private key of $\mathsf{id}_{|k-1}$. Then,

$$d_{\mathsf{id}|k-1} = (xP_2 + r'(\mathsf{id}_1 Q_1 + \cdots + \mathsf{id}_{k-1} Q_{k-1} + P_3), r'P, r'Q_k, r'Q_{k+1}, \ldots, r'Q_h)$$
$$= d_0', d_1', d_k', d_{k+1}', \ldots, d_h' \ (\text{say}).$$

The entity $\mathsf{id}_{|k-1}$ chooses a random $\tilde{r} \in \mathbb{Z}_p^*$ and computes the private key of id as follows:

$$d_0 = d_0' + \mathsf{id}_k d_k' + \tilde{r}(\mathsf{id}_1 Q_1 + \cdots + \mathsf{id}_{k-1} Q_{k-1} + \mathsf{id}_k Q_k + P_3)$$

and the rest of the private key as

$$d_1 = d_1' + \tilde{r}P, d_{k+1} = d_{k+1}' + \tilde{r}Q_{k+1}, \ldots, d_h = d_h' + \tilde{r}Q_h.$$

It is easy to see that this forms a proper private key for id and note that the size of the private key shrinks as we move down the hierarchy.

Encrypt: To encrypt $M \in G_T$ under the identity $\mathsf{id} = (\mathsf{id}_1, \ldots, \mathsf{id}_k) \in (\mathbb{Z}_p^*)^k$, pick a random $s \in \mathbb{Z}_p$ and output

$$\mathsf{CT} = (e(P_1, P_2)^s \times M, sP, s(\mathsf{id}_1 Q_1 + \ldots + \mathsf{id}_k Q_k + P_3)).$$

Decrypt: To decrypt $\mathsf{CT} = (A, B, C)$ using $d_{\mathsf{id}} = (a_0, a_1, b_{k+1}, \ldots, b_h)$, compute

$$A \times \frac{e(a_1, C)}{e(B, a_0)} = M.$$

The correctness of decryption follows from the following computation.

$$\frac{e(a_1, C)}{e(B, a_0)} = \frac{e(rP, s(\mathsf{id}_1 Q_1 + \cdots + \mathsf{id}_k Q_k + P_3))}{e(sP, xP_2 + r(\mathsf{id}_1 Q_1 + \cdots + \mathsf{id}_k Q_k + P_3))}$$
$$= \frac{1}{e(xP, P_2)^s} = \frac{1}{e(P_1, P_2)^s}.$$

Note that, apart from the masked message, the ciphertext in BBG-HIBE consists of only two elements of G irrespective of the number of components in the corresponding identity. In other HIBEs, the length of the ciphertext is proportional to the length of the identity tuple. The BBG-HIBE offers new and important applications for constructing other cryptographic primitives like forward secure encryption [53] and broadcast encryption [139, 78].

Security of BBG-HIBE against adaptive chosen plaintext attack is proved in the selective-ID model under the h-DBDHE assumption [35]. Later [36], the authors showed that some of the extra elements in the h-DBDHE assumption are not required and it is possible to base the security on the hardness of the h-wDBDHI* (or h-wDBDHI) problem. The security reduction uses an algebraic technique similar to that of BB-HIBE. We do not provide the details of this reduction. Interested readers can consult [35, 36] for the complete security argument.

Composite HIBE. Boneh, Boyen and Goh also suggested a "product" construction of the constant size ciphertext BBG-HIBE and BB-HIBE [35]. In case of BBG-HIBE the private key size decreases with the increase in identity level. While in case of BB-HIBE the private key size increases with the height of an identity. Utilizing the algebraic similarities of both the systems they construct a composite scheme where the inner HIBE is the BBG-HIBE and the outer HIBE is the BB-HIBE. The composite scheme allows a trade-off between the ciphertext size and the private key size.

5.3 Interpreting Security Models

The full security model is currently accepted as the most general security model for (H)IBE. In other words, it provides any entity (having any particular identity) in the (H)IBE with the most satisfactory security assurance that the entity can hope for. The notion of security based on an appropriate adversarial game is adapted from the corresponding notion for public key encryption and the security assurance provided in that setting also applies to the (H)IBE setting. The additional consideration is that of identity and the key extraction queries to $\mathcal{O}_k$. We may consider the identity present during the challenge stage to be a target identity. In other words, the adversary wishes to break the security of the corresponding entity. In the full model, the target identity can be any identity, with the usual restriction that the adversary does not know the private key corresponding to this identity or one of its prefixes (in case of HIBE).

From the viewpoint of an individual entity **e** in the (H)IBE structure, the adversary's behavior appears to be the following. The adversary can possibly corrupt any entity in the structure, but as long as it is not able to corrupt that particular entity **e** or one of its ancestors, then it will not be able to succeed in an attack where the target identity is that of **e**. In other words, obtaining the private keys corresponding to the other identities does not help the adversary. Intuitively, that is the maximum protection that any entity **e** can expect from the system.

Let us reflect on the sID model. In this model, the adversary commits to an identity even before the set-up of the (H)IBE is done. The actual set-up can depend on the identity in question. Now consider the security assurance obtained by an individual entity **e**. Entity **e** can be convinced that if the adversary had targeted its identity and then the HIBE structure was set-up, in that case the adversary will not be successful in attacking it. Alternatively, **e** can be convinced that the (H)IBE structure can be set-up so as to protect it. Inherently, the sID model assures that the (H)IBE structure can be set-up to protect any identity, but only one.

Suppose that a (H)IBE structure which is secure in the sID model has already been set-up. It has possibly been set-up to protect one particular identity. The question now is what protection does it offer to entities with other identities? The model does not assure that other identities will be protected. Of course, this does not mean that other identities are vulnerable. The model simply does not say anything about these identities.

It has been observed in [32] that there is a generic conversion from an IBE protocol secure in the selective-ID model to a protocol secure in the adaptive-ID model. However, the resulting protocol suffers from a huge security degradation. If the hash function used to map identities to $\mathbb{Z}_p$ is modeled as a random oracle then the degradation is of the order of q_H (the number of hash queries). Without the random oracle assumption this degradation is of the order of 2^ℓ, where identities are ℓ-bit strings. This further indicates the inadequacy of the selective-ID model. It is also formally established [13] that the strongest notion of security in the selective-ID model does not even imply the weakest notion of security in the adaptive-ID setting.

5.4 Conclusion

This chapter described in details the construction and security argument of BB-(H)IBE in the selective-identity model and also the construction of BBG-HIBE. Apart from the intrinsic importance of these schemes, the algebraic techniques used in the construction and the security reduction have been widely used later. In particular, the ideas introduced in the context of BB-IBE played a pivotal role in the construction of IBE schemes which are secure against adaptive-identity attacks.

Chapter 6
Security Against Adaptive Chosen Ciphertext Attacks

The Boneh-Franklin IBE scheme has two variations. The first variation, called Ba-sicIdent, is secure against adversaries that are *not* allowed to make decryption queries. This construction is modified to obtain the second variation, called Ful-lIdent, which is secure against adversaries that are allowed to make decryption queries. The method of conversion uses a transformation originally proposed by Fujisaki and Okamoto in the context of PKE schemes. This conversion works in the random oracle model.

As a general approach, it is of interest to be able to first obtain a CPA-secure IBE scheme and then convert it to a CCA-secure IBE scheme. This two-stage strategy makes it easier in the security argument to handle key-extraction queries as part of proving CPA-security and then handle decryption queries as part of proving CCA-security. It would actually be nice to have a generic method (without random oracle) of converting a CPA-secure IBE scheme to a CCA-secure IBE scheme. But, till date, no such method is known.

A related result, however, is known. Canetti, Halevi and Katz [54] introduced a generic method for converting any $(h + 1)$-level CPA-secure HIBE to an h-level CCA-secure HIBE. The transformation does not require the random oracle assumption. Setting $h = 0$ gives a conversion from a CPA-secure IBE scheme to a CCA-secure PKE scheme. This transformation uses a strongly unforgeable one-time signature scheme. Boneh and Katz suggested a modification [44] where the one-time signature scheme is replaced by a message authentication code (MAC), resulting in increased efficiency of the transformation. Here we describe the details of the signature-based approach.

Other methods for attaining CCA-security have been proposed. Boyen, Mei and Waters [49] provide a non-generic transformation. Though non-generic, the method is applicable to a large number of pairing based schemes. Variations of the BMW technique have been used for specific (H)IBE schemes [154, 121]. In this chapter, we describe the basic BMW technique.

Gentry [91] proposed an IBE scheme and built it in two stages – the first stage achieves CPA-security while the second stage achieves CCA-security. The technique used by Gentry to achieve CCA-security is based on the proof technique used by Cramer and Shoup [74] to build a CCA-secure PKE scheme. Later in this book, we provide the details of Gentry's CPA-secure IBE scheme. The CCA-security of the second scheme is not discussed.

6.1 A High Level Description

Before getting into the details of any particular transformation, let us briefly consider the problem at a high level. A proof for a CPA-secure IBE scheme requires a mechanism to handle key-extraction requests. In other words, in the security game, if the adversary makes a key extraction query on a particular identity, then, in the proof, it is possible to construct a proper decryption key (subject to certain conditions) and return it to the adversary. So, if the scheme is CPA-secure, then we are assured that its key extraction oracle can be efficiently simulated. However, the scheme may become completely insecure if the adversary is allowed to obtain decryption of a ciphertext of its choice. This is the case with both BasicIdent and Boneh-Boyen IBE discussed earlier in this book.

The above discussion suggests that some additional cryptographic measure is necessary to allow the decryption queries. A decryption query provides an identity and a ciphertext. So, a naive way to answer the query would be to generate a decryption key for the queried identity and use it to decrypt the ciphertext and answer the adversary. This strategy, however, may run into problem if the identity provided as part of the decryption query happens to be the challenge identity, i.e., the identity on which the adversary asked for a challenge ciphertext. Recall that in the security reduction of both BasicIdent and Boneh-Boyen (H)IBE, the simulator cannot generate a valid private key of the challenge identity. Even if the simulator can generate a valid private key for the challenge identity, as is the case with Boneh-Katz-Wang IBE discussed in Chapter 4, the proof (and hence the protocol) may still run into problem. Given the challenge ciphertext, the adversary may apply some trivial transformation on it to generate another valid ciphertext whose decryption will give the adversary some non-trivial information about the challenge. In order to achieve CCA-security, one needs the assurance that all such attempts of the adversary will fail.

6.2 Canetti-Halevi-Katz Transformation

Canetti, Halevi and Katz [54] showed that the problem can be resolved in the context of (H)IBE by using a cryptographic primitive called one-time signature. So we begin with the notion of one-time signatures.

6.2.1 One-Time Signatures

A signature scheme is defined by three probabilistic polynomial time algorithms as follows:

Key-Generation: On input the security parameter 1^κ, this probabilistic polynomial time algorithm outputs a pair of signing key (sk) and verification key (vk).

Sign: This algorithm takes as input a signing key sk and a message M from the appropriate message space $\mathcal{M}$ and outputs a signature σ.

Verify: This is a deterministic algorithm which on input a verification key vk, a message M and a signature σ on M outputs accept or reject depending on whether σ is a valid signature on M or not.

As the name implies, a one-time signature means the signing key is used only once to sign a single message. A signature scheme (Key-Generation, Sign, Verify) is a strong, one-time signature scheme if the success probability of any probabilistic polynomial time adversary $\mathcal{A}$ is negligible in the following game.

1. Key-Generation(1^κ) outputs (vk, sk). The adversary $\mathcal{A}$ is given vk.
2. $\mathcal{A}(1^\kappa, vk)$ may take one of the following actions:

   a. $\mathcal{A}$ outputs a message M and in return is given a signature of M under the signing key sk, i.e., $\sigma \leftarrow \text{Sign}_{sk}(M)$. Then $\mathcal{A}$ outputs a pair (M^*, σ^*).
   b. $\mathcal{A}$ outputs a pair (M^*, σ^*) and halts. In this case (M, σ) is undefined, i.e., the adversary outputs a possible forgery without even seeing a single valid message-signature pair.

$\mathcal{A}$ succeeds in the game if σ^* is a proper signature of M^* under the verification key vk, i.e., $\text{Verify}_{vk}(M^*, \sigma^*) = \text{accept}$ but $(M^*, \sigma^*) \neq (M, \sigma)$. Note that, $\mathcal{A}$ may succeed even if $M^* = M$, which is the reason to call the scheme a strong one-time signature.

6.2.2 The Transformation

Let $\mathcal{H}' = (\text{Set-Up}', \text{Key-Gen}', \text{Encrypt}', \text{Decrypt}')$ be the description of an $(h+1)$-HIBE for arbitrary $h \geq 1$ handling $(n+1)$-bit identities. Let Sig = (Key-Generation, Sign, Verify) be a signature scheme which outputs an n-bit verification key. If $\mathcal{H}'$ is secure in the sense IND-ID-CPA and Sig is a strong one-time signature scheme, then one can construct an h-HIBE $\mathcal{H}$ secure in the sense IND-ID-CCA that handles n-bit identities.

Given an identity tuple id $= (\text{id}_1, \ldots, \text{id}_j) \in (\{0,1\}^n)^j$ of $\mathcal{H}$ we map it to an identity tuple of $\mathcal{H}'$ as

$$\text{Encode(id)} = (0\text{id}_1, \ldots, 0\text{id}_j) \in (\{0,1\}^{n+1})^j$$

and $\mathsf{Encode}(\varepsilon) = \varepsilon$, i.e the null string is mapped to itself. Let $\hat{\mathsf{id}} = \mathsf{Encode}(\mathsf{id})$. The HIBE $\mathcal{H}$ is constructed in such a way that the private key d_{id} of an identity tuple id in $\mathcal{H}$ is equal to the private key $d'_{\hat{\mathsf{id}}}$ of $\hat{\mathsf{id}}$ in $\mathcal{H}'$.

Construction of $\mathcal{H}$

Set-Up: Same as the Set-Up algorithm of $\mathcal{H}'$. The master key of $\mathcal{H}'$, $\mathsf{msk}_{\mathcal{H}'}$ is the master key, $\mathsf{msk}_{\mathcal{H}}$ of $\mathcal{H}$. Similarly, the public parameter of $\mathcal{H}'$, PP becomes the public parameter of $\mathcal{H}$.

Key-Gen: Let d_{id} be the private key of id. To derive the private key of $(\mathsf{id}, \mathsf{v})$ first obtain $\hat{\mathsf{id}} = \mathsf{Encode}(\mathsf{id})$ and $\hat{\mathsf{v}} = \mathsf{Encode}(\mathsf{v})$. Run $\mathsf{Key\text{-}Gen}'_{d_{\mathsf{id}}}(\hat{\mathsf{id}}, \hat{\mathsf{v}})$ and output the result as $d_{\mathsf{id},\mathsf{v}}$.

In the case of IBE or the first level of HIBE, $\mathsf{id} = \varepsilon$, so the master secret of PKG (call it d_ε) is used to generate the private key of the identity v. Since the master secret of $\mathcal{H}$ is same as the master secret of $\mathcal{H}'$, we have $d_{\mathsf{v}} = d'_{\hat{\mathsf{v}}}$. Proceeding this way we see that for any identity tuple (id,v) in $\mathcal{H}$, $d_{\mathsf{id},\mathsf{v}} = d'_{\hat{\mathsf{id}},\hat{\mathsf{v}}}$, given $d_{\mathsf{id}} = d'_{\hat{\mathsf{id}}}$.

Encrypt: To encrypt a message M to an identity tuple id, run the key generation algorithm of Sig, $\mathsf{Key\text{-}Generation}(1^\kappa)$ to obtain (vk, sk). Let $\hat{\mathsf{id}} = \mathsf{Encode}(\mathsf{id}), (1vk)$, compute $C = \mathsf{Encrypt}'_{\mathsf{PP}}(\hat{\mathsf{id}}, M)$ and $\sigma = \mathsf{Sign}_{sk}(C)$. The ciphertext is the tuple $\langle vk, C, \sigma \rangle$.

Note that the n-bit verification key is padded with 1 to get an $(n+1)$-bit "identity" whereas the Encode function pads an identity by 0. This difference plays a crucial role in the simulation.

Decrypt: Given the ciphertext $\langle vk, C, \sigma \rangle$ encrypted under id and the corresponding private key d_{id}, first check whether $\mathsf{Verify}_{vk}(C, \sigma) = \mathsf{accept}$. If not reject the ciphertext. Otherwise, let $\mathsf{id}' = \mathsf{Encode}(\mathsf{id})$ and run $\mathsf{Key\text{-}Gen}'_{d_{\mathsf{id}}}(\mathsf{id}', (1vk))$ to generate the private key $d* = d'_{\hat{\mathsf{id}}}$ (recall that $\hat{\mathsf{id}} = \mathsf{Encode}(\mathsf{id}), (1vk)$). Then output $M = \mathsf{Decrypt}'_{d*}(\hat{\mathsf{id}}, C)$.

6.2.3 Security

Given an identity tuple $\mathsf{id} = (\mathsf{id}_1, \dots, \mathsf{id}_j)$, $j \leq h$ in $\mathcal{H}$, the sender encrypts the message M to a $(j+1)$ level identity $\hat{\mathsf{id}} = (\mathsf{Encode}(\mathsf{id}), (1vk))$ of $\mathcal{H}'$ where vk is the verification key of the underlying one-time signature scheme. The receiver having identity id can derive the private key of $\hat{\mathsf{id}}$ in $\mathcal{H}'$ from the private key d_{id} in $\mathcal{H}$. This is possible because, as we have seen, the private key of id in $\mathcal{H}$ is same as the private key of $\mathsf{Encode}(\mathsf{id})$ in $\mathcal{H}'$.

Use of a strong one-time signature scheme ensures that the adversary will not be able to modify the challenge ciphertext to form another valid ciphertext. On the other hand, use of a CPA-secure $(h+1)$ level HIBE and encoding the verification key of the signature scheme in a different way than a normal identity component ensure that a proper decryption key can be generated for any ciphertext encrypted under the target identity. This intuitive idea is formalized into a reductionist argument as discussed below.

For simplicity, we will assume that the probability of forging a one-time signature, $\Pr[\text{Forge}]$ is negligible. A more precise argument can be given to show an upper bound on the advantage of breaking the CCA-security of the constructed HIBE in terms of the probability of forging a signature and the advantage of breaking the CPA-secure HIBE.

Under the assumption that the probability of forging a signature is negligible, the following argument shows that an IND-ID-CPA adversary $\mathscr{A}'$ against $\mathscr{H}'$ can be used to construct an IND-ID-CCA adversary $\mathscr{A}$ against $\mathscr{H}$. Note that we are considering the adaptive-ID model and the argument can be easily modified for the selective-ID model.

1. $\mathscr{A}'$ obtains the public parameter PP from its challenger, which it relays to $\mathscr{A}$.
2. In Phase 1, whenever $\mathscr{A}$ asks for the private key of an identity id, $\mathscr{A}'$ asks its challenger for the private key $d_{\hat{\text{id}}}$ where $\hat{\text{id}} = \text{Encode}(\text{id})$ and returns it to $\mathscr{A}$.
3. In Phase 1, on a decryption query of the form $(\text{id}, \langle vk, C, \sigma \rangle)$ from $\mathscr{A}$, $\mathscr{A}'$ first checks whether $\text{Verify}_{vk}(C, \sigma) = \text{accept}$. If not $\mathscr{A}'$ returns reject. Otherwise it asks its challenger for the private key of $(\hat{\text{id}}, 1vk)$ where $\hat{\text{id}} = \text{Encode}(\text{id})$ and uses this private key to decrypt C and returns the resulting plaintext to $\mathscr{A}$.
4. In the challenge stage, $\mathscr{A}$ outputs two messages M_0, M_1 and a target identity tuple $\text{id}^* = \langle \text{id}_1^*, \ldots, \text{id}_j^* \rangle$, $j \leq h$. As per the rule of the game, the private key of id^* or any of its prefix was not revealed in Phase 1. $\mathscr{A}'$ first runs the key generation algorithm Key-Generation of Sig to generate (vk^*, sk^*). It outputs the same messages M_0, M_1 and $(\text{Encode}(\text{id}^*), (1vk^*))$ as its target identity. In response, it receives a challenge ciphertext C^*. Now $\mathscr{A}'$ computes $\sigma^* = \text{Sign}_{sk^*}(C^*)$ and returns the ciphertext $\langle vk^*, C^*, \sigma^* \rangle$ to $\mathscr{A}$.
5. In phase 2, $\mathscr{A}$ makes additional decryption queries and private key extraction queries with the usual restriction that it cannot ask for the private key of id^* or any of its prefix and a decryption of the challenge ciphertext under id^* or any of its prefix. The key extraction queries are answered as in Phase 1. For a decryption query of the form $(\text{id}, vk, C, \sigma)$ from $\mathscr{A}$, $\mathscr{A}'$ takes the following action:

    a. If $\text{id} = \text{id}^*$ and $vk = vk^*$, return reject. (The security of the strong one-time signature scheme ensures that the adversary cannot generate a valid ciphertext in this case.)
    b. If $\text{id} \neq \text{id}^*$ or if $\text{id} = \text{id}^*$ but $vk \neq vk^*$, then $\mathscr{A}'$ sets $\hat{\text{id}} = \text{Encode}(\text{id})$ and requests its challenger for the private key of $(\hat{\text{id}}, (1vk))$. It decrypts the ciphertext using this private key and returns the result to $\mathscr{A}$.

6. Finally $\mathscr{A}$ outputs its guess γ'. The same γ' is output by $\mathscr{A}'$.

In the above simulation, $\mathscr{A}'$ poses as a real challenger for $\mathscr{A}$. Since we have assumed that the probability of forging a signature is negligible, the advantage of $\mathscr{A}$ against $\mathscr{H}$ translates into the advantage of $\mathscr{A}'$ against $\mathscr{H}'$.

Based on this generic transformation, any CPA-secure 2-level HIBE can be used to construct a CCA-secure IBE. More generally, a CPA-secure $(h+1)$-HIBE gives a CCA-secure h-HIBE. Hence, protocol designers can concentrate on constructing protocols that achieve CPA-security (be it in the full model or the selective-ID model) without random oracles and then apply this transformation to achieve CCA-security. Protocols such as the Boneh-Boyen HIBE and the Boneh-Boyen-Goh HIBE described in Chapter 5 accomplish this in the selective-ID model, while the Boneh-Boyen IBE and Waters IBE to be described in Chapter 7 accomplish this in the full model.

6.3 The Boyen-Mei-Waters Transformation

Boyen, Mei and Waters [49] provided a non-generic method to convert a CPA-secure IBE scheme to a CCA-secure PKE. The method also applies for the case of (H)IBE. Their construction is non-generic due to the fact that it relies on the use of pairings. But, the idea they introduce is general enough to be applied to a large number of pairing based IBE schemes.

Here we describe their idea as it applies to the Boneh-Boyen HIBE scheme described in Chapter 5. For the ease of understanding, we will only describe the conversion of a CPA-secure IBE scheme to a CCA-secure PKE scheme. To this end, we first briefly describe the IBE version of the BB-HIBE.

BB-IBE.

Set-Up. The setting is of Type-I pairing $e : G \times G \rightarrow G_T$ where $G = \langle P \rangle$. A random $x \in \mathbb{Z}_p$ is chosen and P_1 is set to be equal to xP. Further two other random points P_2 and Q are chosen from G. The public parameters consist of the tuple (P, P_1, P_2, Q) while the master secret key is xP_2.

Identities are assumed to be elements of $\mathbb{Z}_p$. Arbitrary bit strings are hashed into $\mathbb{Z}_p$ using a collision resistant hash function. Define a public function $F : \mathbb{Z}_p \rightarrow G$ as $F(\alpha) = \alpha P_1 + Q$.

Key Generation. Given an identity $\text{id} \in \mathbb{Z}_p$, a decryption key d_{id} is generated as follows. Choose a random $r \in \mathbb{Z}_p$ and compute $d_{\text{id}} = (d_0, d_1) = (xP_2 + rF(\text{id}), rP)$.

Encrypt. The encryption of a message M to an identity id produces a ciphertext (A, B, C) where $A \in G_T$ and $B, C \in G$. These are computed as follows. A random element $s \in \mathbb{Z}_p$ is chosen and (A, B, C) is set to be equal to $(e(P_1, P_2)^s \times M, sP, sF(\text{id}))$.

Decrypt. Given a ciphertext (A, B, C) and an identity id along with a decryption key $d_{\text{id}} = (d_0, d_1)$, the message M is recovered as

$$A \times e(C, d_1) \times e(-B, d_0) = M \times e(P_1, P_2)^s \times e(sF(\text{id}), rP)) \times e(-sP, xP_2 + rF(\text{id}))$$

$$= M \times e(P_1, P_2)^s \times e(-sP, xP_2)$$
$$= M.$$

The basic intuition behind the conversion of this CPA-secure IBE scheme to a CCA-secure PKE scheme is the following. Suppose the user Alice wishes to obtain a public and private key pair. She plays the role of the PKG where the master secret key of the PKG is now the secret key of Alice and the public parameter constitutes her public key. This way the Set-Up algorithm of BB-IBE becomes the key generation algorithm for the PKE scheme. The main novelty comes in the Encrypt algorithm. To encrypt a message to Alice in the PKE scheme, Bob first chooses a random $s \in \mathbb{Z}_p^*$ and computes sP. This is an element of G and constitutes the second element (B) of the ciphertext in BB-IBE. Next a publicly known injective embedding (or a collision resistant hash function) is used to map sP into an element v in $\mathbb{Z}_p$. Now Bob generates the third component (C) of the BB-IBE ciphertext using this v as the "identity". So, effectively, the "identity" used to generate the ciphertext is itself generated dynamically from the second component of the ciphertext and provides the crucial binding for the randomizer s. Note that the second component in the BB-IBE *does not* depend on the identity and that is why this technique can be applied. Bob now masks the message with $e(P_1, P_2)^s$ to generate the first component (A) of BB-IBE ciphertext. Given the ciphertext, Alice first uses the same injective embedding on sP to derive v. Next, she runs the Key-Gen algorithm of BB-IBE on this "identity" v to obtain the corresponding private key. Using this private key she now runs the BB-IBE Decrypt algorithm on the ciphertext (A, B, C) to recover the message.

Though this is the basic idea, the actual conversion provides a CCA-secure key encapsulation mechanism. This can then be combined with a secure data encapsulation mechanism to obtain a PKE scheme. Also, some more refinements can be introduced. These result in improving the efficiencies of the encryption and decryption algorithms. The details of the KEM are as follows. The original paper of Boyen, Mei and Waters described the protocol in the setting of asymmetric pairing. For simplicity we use the setting of symmetric pairing.

BMW-PKE.

Set-Up. Let $e : G \times G \to G_T$ be a symmetric pairing setting and $G = \langle P \rangle$. Let H be an injective encoding from G to $\mathbb{Z}_p$. Choose y_1 and y_2 randomly from $\mathbb{Z}_p$ and P_1 randomly from G. Compute $\xi = e(P, P_1)$, $U_1 = y_1 P$ and $U_2 = y_2 P$. The public key of the user is (H, ξ, U_1, U_2) and the secret key is (P_1, y_1, y_2).

Encapsulate. A random t is chosen from $\mathbb{Z}_p$ and the session key is set to be $\xi^t = e(P, P_1)^t$. The encapsulation of this key is obtained as follows. First compute $B = tP$, then apply the encoding H to B to obtain $w = H(B)$. This w is an element of $\mathbb{Z}_p$ and is the (dynamic) "identity" of the BB-IBE scheme. Compute $C = tU_1 + twU_2$. The encapsulation of the session key is (B, C).

Decapsulate. Given an encapsulation (B,C), and the private key (P_1, y_1, y_2), the session key is re-constructed as follows. First compute $w = H(B)$ which is the "identity" of the BB-IBE scheme used to encapsulate the session key. Next compute $w' = y_1 + wy_2 \pmod{p}$ and check whether $w'B$ equals C. If this equality does not hold, then the ciphertext is not well-formed and the decapsulation algorithm returns $\perp$. If the equality holds, then the session key is obtained as $e(B, P_1)$.

For a properly generated encapsulation key, the validity check will be successful. This is so, because $w'B = (y_1 + wy_2)tP = ty_1 P + twy_2 P = tU_1 + twU_2 = C$. The session key is also correctly reconstructed as $e(B, P_1) = e(tP, P_1) = e(P, P_1)^t = \xi^t$.

Note that in the above description, the component C of the encapsulation is not used for actual reconstruction of the session key. It is used to ensure that the encapsulation itself is well formed. We provide a brief description of the security argument.

In a KEM, there is no message. The adversary for a CCA-secure KEM interacts with the decapsulation oracle during the query stages. For the challenge stage, a session key is generated and is properly encapsulated. This encapsulation is provided to the adversary. Along with the encapsulation, the adversary is provided either the proper session key or a random element from the set of all possible session keys which is independent of the proper session key and both the options are equiprobable. In the guess stage, the adversary outputs its guess of which of the two options have been used.

Coming to the BMW-KEM, the challenge encapsulation is of the form (B^*, C^*), where $B^* = t^*P$. The first thing to note is that this t^* does not depend on the adversary's input. Correspondingly, in the security game, the challenger can choose t^* during the set-up phase and compute $B^* = t^*P$. Once this is done, the challenger also fixes the "challenge identity" $w^* = H(B^*)$. This is the crucial point which indicates that the selective-identity security of the BB-IBE suffices to obtain a CCA-secure PKE.

The challenger is given a DBDH tuple (P, aP, bP, cP, Z) where Z is either $e(P,P)^{abc}$ or Z is a random element of G_T. Note that none of a, b or c is known to the challenger. So, the set-up and answering of the decapsulation queries will have to be done without this knowledge. The challenger sets $w^* = H(cP)$ where H is an injective embedding. In other words, the challenger effectively uses the unknown c as the randomizer t^* for the challenge encapsulation. Next, the challenger chooses random y_1 and y_2 from $\mathbb{Z}_p$ and sets $U_1 = -w^*bP + y_1 P$ and $U_2 = bP + y_2 P$. Further ξ is set to be equal to $e(aP, bP) = e(P, abP)$, i.e., P_1 of the scheme is set to be equal to abP. The public key is declared to be (H, ξ, U_1, U_2) as required, while the secret key is (P_1, y_1, y_2). Note that, the challenger does not actually know $P_1 = abP$.

A decapsulation query is of the type (B, C) where $B = tP$ and $C = t(U_1 + wU_2)$. Such a query is handled by the challenger in two steps. In the first step, it computes $w = H(B) = H(tP)$ as the "identity" to which the implicit encryption of the BB-IBE has been done. Since w, y_1 and y_2 are known to the challenger, it can easily verify the well-formedness of the query. Once this is verified, it proceeds to generate a decryption key for w using the technique for simulating the key extraction queries in the BB-IBE scheme. More specifically, it chooses a random $r \in \mathbb{Z}_p$ and computes (d_0, d_1) in the following manner.

$$d_0 = r(U_1 + wU_2) - \frac{1}{w - w^*}(y_1 + wy_2)aP$$

$$= abP - \frac{w - w^*}{w - w^*}abP - \frac{1}{w - w^*}(y_1 + wy_2)aP + r(U_1 + wU_2)$$

$$= abP - \frac{a}{w - w^*}((w - w^*)bP + (y_1 + wy_2)P) + r(U_1 + wU_2)$$

$$= abP - \frac{a}{w - w^*}(-w^*bP + y_1 P + w(bP + y_2 P)) + r(U_1 + wU_2)$$

$$= abP - \frac{a}{w - w^*}(U_1 + wU_2) + r(U_1 + wU_2)$$

$$= abP + \left(r - \frac{a}{w - w^*}\right)(U_1 + wU_2);$$

$$d_1 = rP - \frac{1}{w - w^*}P_1$$

$$= \left(r - \frac{a}{w - w^*}\right)P.$$

Note that the crucial point is to use the beautiful algebraic technique introduced by Boyen and Boyen [32] for simulating key extraction queries in BB-HIBE discussed in Chapter 5. The above computation holds only if $w \neq w^*$. Since w^* equals $H(cP)$ and c is a random element of $\mathbb{Z}_p$, w^* is also a random element of $\mathbb{Z}_p$. So the probability that w equals w^* is $1/p$. For a total of q decapsulation queries, this accounts for an additive security degradation of q/p. Having generated (d_0, d_1), the challenger proceeds to obtain the session key for (B, C) as follows. Let $r' = r - a/(w - w^*)$.

$$\frac{e(d_0, B)}{e(C, d_1)} = \frac{e(abP + r'(U_1 + wU_2), tP)}{e(t(U_1 + wU_2), r'P)}$$

$$= e(abP, tP) \times \frac{e(r'(U_1 + wU_2), tP)}{e(t(U_1 + wU_2), r'P)}$$

$$= e(P, abP)^t$$

$$= e(P, P_1)^t$$

$$= \xi^t$$

Thus, the challenger can answer the decapsulation query.

For the challenge ciphertext, we have $w^* = H(cP)$ and the challenger returns (B^*, C^*) and Z where $B^* = cP$ and C^* is computed to be

$$C^* = (y_1 + w^* y_2)cP$$

$$= c((w^* - w^*)bP + y_1 P + w^* y_2 P)$$

$$= c(-w^* bP + y_1 P + w^*(bP + y_2 P))$$

$$= c(U_1 + w^* U_2).$$

This shows that a proper encapsulation is provided to the adversary. If Z is $e(P, P)^{abc}$, then along with the encapsulation the adversary gets the proper session key. On the other hand, if Z is random then it is independent of (B^*, C^*) and provides no information to the adversary. If the adversary is able to correctly guess which of the two cases has occurred, then the challenger is able to solve the DBDH problem.

6.4 Conclusion

This chapter describes the techniques of how to achieve CCA-security for (H)IBE. We discussed the generic method of Canetti, Halevi and Katz (CHK) to convert a CPA-secure HIBE into a CCA-secure HIBE. The efficiency of this technique was later improved by Boneh and Katz. We also described the Boyen, Mei and Waters (BMW) technique of converting an IBE which is CPA-secure in the selective-ID model to a CCA secure KEM. The BMW technique is applicable for many pairing based (H)IBE schemes and in the next chapter we will detail this for an HIBE in the adaptive-ID model. HIBE to a CCA-secure

Chapter 7
IBE in Adaptive-Identity Model Without Random Oracles

In the previous chapters, we have seen several IBE schemes and their extension to HIBE. As mentioned in Chapter 2, the security model for IBE schemes was introduced by Boneh and Franklin and allows the adversary to make both adaptive-identity and adaptive-chosen ciphertext attacks. So, a natural goal is to obtain schemes which can be proved secure in this model under the assumption that some computational problem is hard to solve.

In a sense, this was already achieved by the IBE scheme proposed by Boneh and Franklin. But, the proof of security of this scheme assumed that certain hash functions used in the scheme are random oracles. This was not considered satisfactory enough and the search continued for an IBE scheme which can be proved secure without random oracles.

This problem was solved, but, the resulting IBE schemes could not be proved secure in the model proposed by Boneh and Franklin. Specifically, the task of handling adaptive-identity attacks was found to be difficult and the security game was instead weakened to the selective-identity model. Schemes which can be proved secure in this model have been discussed in Chapter 5.

Starting from the work of Boneh and Franklin, it was realised that the two tasks of handling adaptive-identity attacks and adaptive-chosen ciphertext attacks can be separated. It is possible to first build an IBE scheme which is secure against adaptive-identity attacks and then modify it to obtain an IBE scheme which is secure against adaptive-chosen ciphertext attacks. This has been discussed in Chapter 6.

So, the problem now boils down to obtaining an IBE scheme which is secure against adaptive-identity attacks without the use of random oracles in the security reduction. There is also the related question of hardness assumption. The most natural hardness assumption for pairing based schemes is the bilinear Diffie-Hellman assumption and its decisional version, the DBDH assumption. This leads to the basic question of whether it is possible to obtain an IBE scheme which is secure against adaptive-identity attacks without the use of random oracles and whose security can be based on the hardness of the DBDH assumption. This chapter discusses the solutions to this question.

The first solution to this problem was proposed by Boneh and Boyen. Their construction can be seen as a modification of the Boneh-Boyen selective-identity secure (H)IBE scheme discussed in Chapter 5. The core idea is a modification of the ingenious strategy for simulating decryption key queries that had been introduced in the context of selective-identity security.

In the Boneh-Boyen selective-identity secure IBE, the master secret key is an element S of G. Given an identity id, the decryption key is defined to be $S + rV(\text{id})$, where r is chosen randomly from $\mathbb{Z}_p$ and $V(\text{id})$ "hashes" the identity to an element of G. This structure has turned out to be extremely useful in later IBE schemes achieving security against adaptive-identity attacks.

Boneh and Boyen [33] themselves proposed a particular form of $V(\text{id})$ which provided the first construction of an adaptive-identity secure IBE scheme without the use of random oracles. The scheme itself, however, was rather inefficient and was of more theoretical than practical interest. Waters [169] provided a modified form for $V(\text{id})$ which led to efficient encryption and decryption algorithms. One drawback which remained was that Waters IBE scheme required public parameters which were rather large. Independent work by Chatterjee and Sarkar [60] and Naccache [137] addressed this problem. This chapter describes these IBE schemes.

Extension of Waters' IBE scheme to a HIBE was suggested by Waters himself. This required independent sets of group elements for every level of the HIBE. Chatterjee and Sarkar [61] showed that most of the group elements can be reused across the levels of the HIBE. This HIBE scheme is described later in the chapter. We also discuss how to obtain the CCA-secure version of the (H)IBE schemes described here.

7.1 Boneh-Boyen IBE

Identities are elements of $\{0,1\}^w$ and they are mapped to a random n bit string through a hash function H_k. H_k is chosen from a family of hash functions $\{H_k : \{0,1\}^w \to \{0,1\}^n\}_{k \in \mathscr{K}}$, where $\mathscr{K}$ is the key space for the family of hash functions.

Set-Up: Choose an arbitrary generator $P \in G$, pick a random $x \in \mathbb{Z}_p$ and set $P_1 = xP$. Also choose a random element $P_2 \in G$. Construct a random $n \times 2$ matrix $\mathscr{U} = (U_{i,j}) \in G^{n \times 2}$ where each $U_{i,j}$ is uniform in G. Finally pick a random hash function key $k \in \mathscr{K}$. The public parameters are $PP = (P, P_1, P_2, \mathscr{U}, k)$ and the master secret key is $\text{msk} = xP_2$.

Key-Gen: To generate the private key d_{id} for an identity $\text{id} \in \{0,1\}^w$, compute $\vec{a} = H_k(\text{id}) = (a_1, \ldots, a_n) \in \{0,1\}^n$ and pick random $r_1, \ldots, r_n \in \mathbb{Z}_p$. The private key is $d_{\text{id}} = (xP_2 + \sum_{i=1}^n r_i U_{i,a_i}, r_1 P, \ldots, r_n P)$. Note that the private key consists of $n+1$ elements of G.

Encrypt: To encrypt a message $M \in G_T$ for the identity $\text{id} \in \{0,1\}^w$, set $\vec{a} = H_k(\text{id}) = (a_1, \ldots, a_n)$, pick a random $s \in \mathbb{Z}_p$ and compute

$$C = (e(P_1, P_2)^s \times M, sP, sU_{1,a_1}, \ldots, sU_{n,a_n}) \in G_T \times G^{n+1}$$

Decrypt: To decrypt $C = (A, B, C_1, \ldots, C_n)$ using $d_{\mathrm{id}} = (d_0, d_1, \ldots, d_n)$, compute

$$A \times \frac{\prod_{j=1}^n e(C_j, d_j)}{e(B, d_0)} = M$$

It is easy to verify that this gives a proper decryption.

Security We do not provide the complete security reduction for this scheme. Only the basic idea is discussed. At a high level, the construction can be seen to have similarities with the Boneh-Boyen selective-identity secure HIBE scheme described in Chapter 5. Consider a BB-HIBE of maximum depth n where the identity tuples are of the form $\mathrm{id} = (i_1, \ldots, i_n)$, i.e., only full depth identity tuples are allowed. Then what we essentially get is the above construction.

In this case, however, each $i_j \in \{0, 1\}$ whereas in the original BB-HIBE construction, each i_j is an element of $\mathbb{Z}_p$. To tackle this difference, Boneh-Boyen use additional randomisation in the form of the hash function $H_k()$ and the $n \times 2$ matrix $\mathscr{U}$.

The difference between the requirements of proving selective-identity security and adaptive-identity security raises certain problems. In the security reduction, first the challenger $\mathscr{B}$ randomly chooses an n-bit string $\mathrm{id}^c = (i_1^c, \ldots, i_n^c)$. Then $\mathscr{B}$ forms the matrix $\mathscr{U}$ such that it can form the private key for any identity $\mathrm{id} = (i_1, \ldots, i_n)$ only if there is some j, $1 \le j \le n$ such that $i_j = i_j^c$. In contrast, in the challenge phase, it can form a proper encryption only if the challenge identity, $\mathrm{id}^* = (i_1^*, \ldots, i_n^*)$ is the complement string of id^c, i.e., there is no j such that $i_j^* = i_j^c$ for $1 \le j \le n$. The security reduction suffers from a rather large degradation because the simulator can generate a proper encryption only for a single identity.

7.2 Waters IBE

In the Boneh-Boyen IBE scheme described in the previous section, the public parameters, private key and ciphertext sizes are quite large. In the IBE scheme proposed by Waters [169] the sizes of the private key and the ciphertext are drastically reduced.

Set-Up: Identities are defined to be n-bit strings. Choose an arbitrary generator $P \in G$ and a uniform random x from $\mathbb{Z}_p$. Set $P_1 = xP$; choose $P_2 \in G$ at random. Further, choose a random element $U' \in G$ and a random n-length vector $\overrightarrow{U} = \{U_1, \ldots, U_n\}$, whose elements are from G. The master secret is xP_2 whereas the public parameter consists of $(P, P_1, P_2, U', \overrightarrow{U})$.

Key-Gen: Let $\mathrm{id} = (\mathrm{id}_1, \ldots, \mathrm{id}_n) \in \{0, 1\}^n$ be any identity. A secret key for id is generated as follows. Choose a random $r \in \mathbb{Z}_p^*$, then the private key for id is

$$d_{id} = (xP_2 + rV, rP).$$

where

$$V = U' + \sum_{\{i:id_i=1\}} U_i.$$

Encrypt: Any message $M \in G_T$ is encrypted for an identity id as

$$C = (e(P_1, P_2)^t \times M, tP, tV),$$

where t is a random element of $\mathbb{Z}_p$ and V is as defined in **Key-Gen** above.

Decrypt: Let $C = (C_1, C_2, C_3)$ be a ciphertext and id be the corresponding identity. Then we decrypt C using secret key $d_{id} = (d_1, d_2)$ by computing

$$C_1 \times \frac{e(d_2, C_3)}{e(d_1, C_2)}$$

The correctness of the construction can be seen from the following computation.

$$
\begin{aligned}
C_1 \times \frac{e(d_2, C_3)}{e(d_1, C_2)} &= C_1 \times \frac{e(rP, tV)}{e(xP_2 + rV, tP)} \\
&= M \times e(P_1, P_2)^t \times \frac{e(rP, tV)}{e(xP_2, tP)e(rV, tP)} \\
&= M \times e(P_1, P_2)^t \times \frac{e(rP, tV)}{e(P_1, P_2)^t e(rP, tV)} \\
&= M.
\end{aligned}
$$

7.2.1 Security

Security of Waters IBE against chosen plaintext attack is established through the following theorem.

Theorem 7.1. *The Waters IBE protocol described in Section 7.2 is $(\varepsilon_{ibe}, t, q)$-CPA secure assuming that the (ε_{dbdh}, t')-DBDH assumption holds in (G, G_T, e), where*

$$\varepsilon_{ibe} \leq 16nq\varepsilon_{dbdh}$$

where n is the bit-length of identities, q is the maximum number of key extraction queries and $t' = t + O(\tau q)$; τ is the time required for one scalar multiplication in G.

We want to show that the Waters IBE is $(\varepsilon_{ibe}, t, q)$-CPA secure. This is established through a game sequence style of proofs discussed in Chapter 2. We start with the adversarial game defining the CPA-security of the protocol against an adversary $\mathscr{A}$

and then obtain a sequence of games. In each of the games, the simulator chooses a bit γ and the adversary makes a guess γ'. By X_i we will denote the event that the bit γ is equal to the bit γ' in the ith game.

Game 0:

This is the usual adversarial game used in defining CPA-secure IBE. We assume that the adversary's runtime is t and it makes q key extraction queries. Also, we assume that the adversary maximizes the advantage among all adversaries with similar resources. Thus, we have $\varepsilon_{ibe} = \left| \Pr[X_0] - \frac{1}{2} \right|$.

Game 1:

Consider a tuple $(P, P_1 = aP, P_2 = bP, P_3 = cP, Z = e(P,P)^{abc})$ where a, b and c are chosen uniformly and independently at random from $\mathbb{Z}_p$. The simulator is assumed to know the values a, b and c. But, the simulator can setup the protocol as well as answer certain private key queries without the knowledge of these values. Also, for certain challenge identities it can generate the challenge ciphertext without the knowledge of a, b and c. In the following, we show how this can be done. If the simulator cannot answer a key extraction query or generate a challenge without using the knowledge of a, b and c, it sets a flag flg to one. The value of flg is initially set to zero.

Note that the simulator is always able to answer the adversary (with or without using a, b and c). The adversary is provided with proper replies to all its queries and is also provided the proper challenge ciphertext. Thus, irrespective of whether flg is set to one, the adversary's view in Game 1 is same as that in Game 0. Hence, we have $\Pr[X_0] = \Pr[X_1]$.

We next show how to setup the protocol and answer the queries based on the tuple $(P, P_1 = aP, P_2 = bP, P_3 = cP, Z = e(P,P)^{abc})$.

Set-Up: Let $m = 4q$. Choose $x', x_1, \ldots, x_n$ randomly from $\mathbb{Z}_m$; $y', y_1, \ldots, y_n$ randomly from $\mathbb{Z}_p$. Choose k randomly from $\{0, \ldots, n\}$.

Now define $U' = (p - mk + x')P_2 + y'P$ and for $1 \le i \le n$ define $U_i = x_i P_2 + y_i P$. Set the public parameters of IBE to be $(P, P_1, P_2, U', U_1, \ldots, U_n)$. The master secret is $aP_2 = abP$. In its attack, $\mathscr{A}$ will make some queries, which have to be properly answered by the simulator.

Let $v = (v_1, \ldots, v_n)$ be an n-bit string. We define the following functions:

$$
\left.
\begin{aligned}
F(v) &= p - mk + x' + \sum_{i=1}^{n} x_i v_i \\
J(v) &= y' + \sum_{i=1}^{n} y_i v_i \\
L(v) &= x' + \sum_{i=1}^{n} x_i v_i \pmod{m} \\
K(v) &= \begin{cases} 0 & \text{if } L(v) = 0 \\ 1 & \text{otherwise.} \end{cases}
\end{aligned}
\right\}
\tag{7.2.1}
$$

Let F_{min} and F_{max} be the minimum and maximum values of $F(v)$. F_{min} is achieved when k is maximum and x' and the x_i's are all zero. Thus, $F_{min} = p - mn > 0$ (for practical choices of m, n and p). Similarly, F_{max} is achieved when $k = 0$ and x', x_i's and v_i's are equal to their respective maximum values. We get $F_{max} < p + m(n+1) < 2p$ (again for practical choices of m, n and p, $m(n+1) < p$). Consequently, $F(v) \equiv 0 \bmod p$ if and only if $F(v) = p$ which holds if and only if $x' + \sum_{i=1}^{n} x_i v_i = mk$.

Now we describe how the key extraction queries made by $\mathscr{A}$ are answered. The queries can be made in both Phase 1 and Phase 2 of the adversarial game (subject to the usual restrictions). The manner in which they are answered by the simulator is the same in both phases.

Key Extraction Query: Suppose $\mathscr{A}$ makes a key extraction query on the identity $\mathrm{id} = (\mathrm{id}_1, \ldots, \mathrm{id}_n)$. Choose a random r from $\mathbb{Z}_p$. Suppose $K(\mathrm{id}) = 1$; otherwise set flg to 1. In the second case, the simulator uses the value of a to return the proper decryption key $d_{\mathrm{id}} = (aP_2 + rV(\mathrm{id}), rP)$. In the first case, the simulator constructs a decryption key in the following manner.

$$\left.\begin{array}{l} d_0 = -\frac{J(\mathrm{id})}{F(\mathrm{id})}P_1 + r(F(\mathrm{id})P_2 + J(\mathrm{id})P) \\ d_1 = \frac{-1}{F(\mathrm{id})}P_1 + rP \end{array}\right\} \tag{7.2.2}$$

The quantity $d_{\mathrm{id}} = (d_0, d_1)$ is a proper private key corresponding to the identity id. To see this, suppose $r' = r - a/F(\mathrm{id})$, then

$$\begin{aligned} d_0 &= -\frac{J(\mathrm{id})}{F(\mathrm{id})}P_1 + r(F(\mathrm{id})P_2 + J(\mathrm{id})P) \\ &= abP - \frac{F(\mathrm{id})}{F(\mathrm{id})}abP - \frac{J(\mathrm{id})}{F(\mathrm{id}}aP + r(F(\mathrm{id})P_2 + J(\mathrm{id})P) \\ &= aP_2 - \frac{a}{F(\mathrm{id})}(F(\mathrm{id})P_2 + J(\mathrm{id})P) + r(F(\mathrm{id})P_2 + J(\mathrm{id})P) \\ &= aP_2 + r'(F(\mathrm{id})P_2 + J(\mathrm{id})P) \\ &= aP_2 + r'((p - mk + x')P_2 + y'P + \sum_{i=1}^{n}(\mathrm{id}_i(x_iP_2 + y_iP))) \\ &= aP_2 + r'(U' + \sum_{i=1}^{n}\mathrm{id}_iU_i) \end{aligned}$$

and it is easy to see that $d_1 = r'P$. This $d_{\mathrm{id}} = (d_0, d_1)$ is provided to $\mathscr{A}$.

Challenge: Let the challenge identity be $\mathrm{id}^* = (\mathrm{id}_1^*, \ldots, \mathrm{id}_n^*)$, and the (equal length) messages be M_0 and M_1. Choose a random bit γ. We need to have $F(\mathrm{id}^*) \equiv 0 \bmod p$. If this condition does not hold, then set flg to 1. In the second case, the simulator uses the value of c to provide a proper encryption of M_γ to $\mathscr{A}$ by computing $(M_\gamma \times e(P_1, P_2)^c, cP, cV(\mathrm{id}^*))$. In the first case, it constructs a proper encryption of M_γ in the following manner.

$$(M_\gamma \times Z, C_1 = P_3, C_2 = J(\text{id}^*)P_3).$$

We require C_2 to be equal to $cV(\text{id}^*)$. Recall that the definition of $V(\text{id})$ is $V(\text{id}) = U' + \sum_{i=1}^n \text{id}_i U_i$. Using the definition of U' and the U_i's as defined in Set-Up by the simulator, we obtain,

$$
\begin{aligned}
cV(\text{id}^*) &= c(U' + \sum_{i=1}^n \text{id}_i^* U_i) \\
&= c((p - mk + x')P_2 + y'P + \sum_{i=1}^n (\text{id}_i^*(x_i P_2 + y_i P))) \\
&= c(F(\text{id}^*)P_2 + J(\text{id}^*)P) \\
&= J(\text{id}_i^*)cP \\
&= J(\text{id}^*)P_3
\end{aligned}
$$

Here we use the fact, $F(\text{id}^*) \equiv 0 \mod p$.

Guess: The adversary outputs a guess γ' of γ and wins if they are equal.

Game 2:

This is a modification of Game 1 whereby the $Z = e(P,P)^{abc}$ in Game 1 is now replaced by a random element of G_T. This Z is used to mask the message M_γ in the challenge ciphertext. Since Z is random, the first component of the challenge ciphertext is a random element of G_T and provides no information to the adversary about γ. Thus, $\Pr[X_2] = \frac{1}{2}$.

We show that it is possible to construct an algorithm $\mathscr{B}$ for solving the DBDH problem by extending Game 1 and Game 2. The extension of both the games is same and is described as follows. $\mathscr{B}$ takes as input a tuple (P, aP, bP, cP, Z) and sets up the IBE protocol as in Game 1. Also, the key extraction queries are answered and the challenge ciphertext is generated as in Game 1. If at any stage, flg is set to 1, then $\mathscr{B}$ outputs a random bit and aborts. At the end of the game, the adversary outputs the guess γ'. If $\mathscr{B}$ has not aborted up to this stage, then it outputs 1 if $\gamma = \gamma'$; else 0.

If Z is equal to $e(P,P)^{abc}$, then the adversary is playing Game 1 and if Z is a random element of G_T, then the adversary is playing Game 2. The time taken by $\mathscr{B}$ in either Game 1 or 2 is clearly t'.

Suppose $\Pr[\text{flg}_i = 0]$ is the probability that the simulator $\mathscr{B}$ does not abort during Game i, $i = 1, 2$. In order to relate the advantage of $\mathscr{A}$ against the IBE scheme with the advantage of $\mathscr{B}$ to solve the DBDH problem, we need to find a bound on this probability.

Bounds on Probability of Not Abort

Before proceeding further with our analysis we state the following simple result which can be easily verified from elementary probability theory.

Proposition 7.1. *Let X and Y be discrete random variables.*

1. *If* $\Pr[X = c_1 | Y = c_2]$ *is a constant (i.e., the probability does not depend on c_1 and c_2), then X and Y are independent.*
2. *If X and Y are independent and uniformly distributed, then*

$$\Pr[g(X, f_1(Y)) = c_1 | f_1(Y) = c_2, f_2(Y) = c_3] = \Pr[g(X, f_1(Y)) = c_1 | f_1(Y) = c_2]$$

where f_1, f_2 and g are arbitrary functions (with appropriate domains and ranges) and c_1, c_2 and c_3 are arbitrary elements in the respective sets.

Let flg_i denote the random variable flg in Game i, $i = 1, 2$. Suppose $\mathsf{id}^{(1)}, \ldots, \mathsf{id}^{(q)}$ are the identities in the q key extraction queries and id^* is the challenge identity. Let $\mathsf{V} = (\mathsf{id}^*, \mathsf{id}^{(1)}, \ldots, \mathsf{id}^{(q)})$. Let $\mathbf{X}$ be the tuple of random variables consisting of the x_i's and x' used during set-up. Let $\mathbf{Z}$ be the tuple of random variables consisting of the adversary's private random bits; the y_i's and y' used during set-up; and the r's used in answering the key extraction queries. A specific value of $\mathbf{X}$ will be denoted by $\mathbf{x}$; a specific value of $\mathbf{Z}$ will be denoted by $\mathbf{z}$; and a specific value of V will be denoted by $\mathbf{v}$. The following observation is due to Bellare and Ristenpart [23].

Proposition 7.2. *1.* V *is independent of* $\mathbf{X}$.
2. In the i-th game, the event X_i is independent of $\mathbf{X}$.
3. The random variable flg_i is a function $\mathsf{flg}_i \triangleq \mathsf{flg}_i(\mathbf{X}, \mathsf{V})$.

Proof : (1) Fix any value $\mathbf{x}$ of $\mathbf{X}$. Irrespective of this value, the independent and uniform random choices of the y_i's and y' ensure that the public parameters are independent and uniformly distributed points. Similarly, the independent and uniform random choices of the r's ensure that the response to any query is uniform random and independent of other random variables. Similarly, the independent and uniform randomness of c ensures that the challenge ciphertext is independent of $\mathbf{X}$. This is true irrespective of whether flg_i is set to 1 or 0.

The adversary's queries depends on its own random choices (which is independent of $\mathbf{X}$), the distribution of the public parameters, the responses to the queries and the challenge ciphertext. By the above argument, for every fixed value of $\mathbf{X}$, the distribution of these random variables are the same. Hence, for every fixed value of $\mathbf{X}$, the probability that the adversary outputs a particular query sequence is a constant, i.e. $\Pr[\mathsf{V} = \mathbf{v} | \mathbf{X} = \mathbf{x}]$ is a constant. From Proposition 7.1(1), it follows that V is independent of $\mathbf{X}$.

(2) The bit γ is a uniform random bit which is independent of all other quantities. In Games 0 and 1, the adversary's output γ' is a function of its private random choices, the public parameters, the responses to the queries and the challenge ciphertext. So, as argued above, the output γ' is independent of $\mathbf{X}$ and hence the event

$\gamma = \gamma'$ is also independent of $\mathbf{X}$. In Game 2, the bit γ is statistically hidden from the adversary and hence the probability of X_2 is $1/2$ irrespective of the value of $\mathbf{X}$.

(3) The value of flg_i is 0 if all the F-values corresponding to the key extraction queries are non-zero and the F-value for the challenge identity is 0. From the definition of the function F, it follows that this event depends only on $\mathbf{X}$ and V and so the random variable flg_i is a function of these two random variables. $\quad\square$

From the above two propositions we obtain the following result.

Proposition 7.3. $\Pr[\mathsf{flg}_i = 0 | \mathsf{V} = \mathbf{v}, X_i] = \Pr[\mathsf{flg}_i = 0 | \mathsf{V} = \mathbf{v}]$.

We further require the following two independence results in obtaining the required bound.

Proposition 7.4. *Let $L(\cdot)$ be as defined in (7.2.1) and let $\mathsf{id} \in \{0,1\}^n$ be any identity, then*

1. $\Pr[L(\mathsf{id}) = 0] = \dfrac{1}{m}$.

2. Let id' be an identity such that $\mathsf{id}' \neq \mathsf{id}$. Then $\Pr\left[L(\mathsf{id}') = 0 \,|\, L(\mathsf{id}) = 0\right] = \dfrac{1}{m}$.

The probability is over the independent and uniform random choices of x' and $x_1, \ldots, x_n$ from $\mathbb{Z}_m$.

Proof : Recall from (7.2.1) that $L(\mathsf{id}) = x' + \mathsf{id}_1 x_1 + \cdots + \mathsf{id}_n x_n$. Each of the values $x', x_1, \ldots, x_l$ are chosen independently and uniformly at random from $\mathbb{Z}_m$. This ensures that $L(\mathsf{id})$ is also independently and uniformly distributed over $\mathbb{Z}_m$. The first point follows from this observation.

For the second point, since $\mathsf{id} \neq \mathsf{id}'$, there is an $i \in \{1, \ldots, n\}$ such that not both of id_i and id'_i are zeros. Without loss of generality, suppose that id'_i is non-zero. Then the result follows from the independent and uniform randomness of $x', x_1, \ldots, x_n$. $\square$

Proposition 7.5. *For any fixed $\mathbf{v}$, let $\lambda(\mathbf{v}) \stackrel{\Delta}{=} \Pr[\mathsf{flg}(\mathbf{X}, \mathbf{v}) = 0]$*

$$\lambda^- \stackrel{\Delta}{=} \left(1 - \frac{q}{m}\right)\lambda^+ \leq \lambda(\mathbf{v}) \leq \lambda^+ \stackrel{\Delta}{=} \frac{1}{m(n+1)}. \qquad (7.2.3)$$

Proof : For any fixed $\mathbf{v}$, let $\mathsf{ab}(\mathbf{v})$ be the event $\mathsf{flg}(\mathbf{X}, \mathbf{v}) = 1$. For $1 \leq i \leq q$, let E_i denote the event that the simulator does not abort on the ith key extraction query and let C be the event that the simulator does not abort in the challenge stage. We have

$$\Pr[\overline{\mathsf{ab}(\mathbf{v})}] = \Pr\left[\left(\bigwedge_{i=1}^{q} E_i\right) \wedge C\right]$$

$$= \Pr\left[\left(\bigwedge_{i=1}^{q} E_i\right) | C\right] \Pr[C]$$

$$= \left(1 - \Pr\left[\left(\bigvee_{i=1}^{q} \neg E_i\right) | C\right]\right) \Pr[C]$$

$$\geq \left(1 - \sum_{i=1}^{q} \Pr[\neg E_i | C]\right) \Pr[C].$$

We first consider the event C. Suppose the challenge identity is id^*. Event C holds if and only if $F(\mathrm{id}^*) \equiv 0 \bmod p$. Recall that by choice of p, we can assume $F(\mathrm{id}^*) \equiv 0 \bmod p$ if and only if $x' + \sum_{i=1}^{n} x_i \mathrm{id}_i^* = mk$. Hence,

$$\Pr[C] = \Pr\left[\left(x' + \sum_{i=1}^{n} x_i \mathrm{id}_i^* = mk\right)\right]. \tag{7.2.4}$$

For $0 \leq j \leq n$, denote the event $x' + \sum_{i=1}^{n} x_i \mathrm{id}_i^* = mj$ by A_j and the event $k = j$ by B_j. Also, let C_j be the event $A_j \wedge B_j$.

Note that the event $\bigvee_{j=0}^{n} A_j$ is equivalent to the condition $x' + \sum_{i=1}^{n} x_i \mathrm{id}_i^* \equiv 0 \bmod m$ and hence equivalent to the condition $L(\mathrm{id}^*) = 0$. Since k is chosen uniformly at random from the set $\{0, \ldots, n\}$, we have $\Pr[B_j] = 1/(1+n)$ for all j. Also the event B_j is independent of the event A_j. We have

$$\Pr\left[\left(x' + \sum_{i=1}^{n} x_i \mathrm{id}_i^* = mk\right)\right] = \Pr\left[\left(\bigvee_{i=0}^{n} C_i\right)\right]$$

$$= \Pr\left[\bigvee_{i=0}^{n} (A_i \wedge B_i)\right]$$

$$= \sum_{i=0}^{n} \Pr[A_i \wedge B_i]$$

$$= \sum_{i=0}^{n} \Pr[A_i] \times \Pr[B_i]$$

$$= \frac{1}{(1+n)} \sum_{i=0}^{n} \Pr[A_i]$$

$$= \frac{1}{(1+n)} \Pr\left[\bigvee_{i=0}^{n} (A_i)\right]$$

$$= \frac{1}{(1+n)} \Pr[(L(\mathrm{id}^*) = 0)]$$

$$= \frac{1}{m(1+n)}$$

The last equality follows from Proposition 7.4. This shows that $\Pr[C] \leq 1/(m(1+n))$ and so

$$\Pr[\overline{ab(\mathbf{v})}] = \Pr\left[\left(\bigwedge_{i=1}^{q} E_i\right) \wedge C\right]$$

$$\leq \Pr[C] \leq \frac{1}{m(1+n)}.$$

This shows the required upper bound.

To obtain the lower bound, we now turn to bounding $\Pr[\neg E_i | C]$. For simplicity of notation, we will drop the subscript i from E_i and consider the event E that the simulator does not abort on a particular key extraction query on an identity id. By the simulation, the event $\neg E$ implies that $L(\mathsf{id}) = 0$. This holds even when the event is conditioned under C. From Proposition 7.4 we have $\Pr[\neg E | C] = 1/m$.

Substituting this in the bound for $\Pr[\overline{ab(\mathbf{v})}]$ we obtain

$$\Pr[\overline{ab(\mathbf{v})}] \geq \left(1 - \sum_{i=1}^{q} \Pr[\neg E_i | C]\right) \Pr[C].$$

$$\geq \left(1 - \frac{q}{m}\right) \frac{1}{m(n+1)}.$$

This completes the proof of Proposition 7.5. $\square$

Based on the bounds on the probability of not abort, we continue our analysis of the relation between the advantage of $\mathscr{A}$ and $\mathscr{B}$. Let Y_i be the event that the simulator outputs 1 in Game i, $i = 1, 2$. Then, we have

$$|\Pr[Y_1] - \Pr[Y_2]| \leq \varepsilon_{dbdh}.$$

Let ab_i be the event $\mathsf{flg}_i(\mathbf{X}, \mathbf{v}) = 1$, i.e., the event that the simulator aborts in Game i, $i = 1, 2$.

Proposition 7.6. $\varepsilon_{ibe} \leq \dfrac{\varepsilon_{dbdh}}{\left(1 - \frac{q}{m}\right) \lambda^+}.$

Proof :

$$\Pr[Y_i] = \Pr[Y_i \mid ab_i] \Pr[ab_i] + \Pr[Y_i \mid \overline{ab_i}] \Pr[\overline{ab_i}]$$

$$= \frac{1}{2} \Pr[ab_i] + \Pr[X_i \mid \overline{ab_i}] \Pr[\overline{ab_i}]$$

$$= \frac{1}{2} \Pr[ab_i] + \Pr[X_i \wedge \overline{ab_i}]$$

$$= \frac{1}{2} \Pr[ab_i] + \Pr[X_i \wedge \mathsf{flg}_i(\mathbf{X}, \mathsf{V}) = 0]$$

$$= \frac{1}{2} \Pr[ab_i] + \sum_{\mathbf{v}} \Pr[X_i \wedge \mathsf{flg}_i(\mathbf{X}, \mathsf{V}) = 0 \wedge \mathsf{V} = \mathbf{v}]$$

$$= \frac{1}{2} \Pr[ab_i] + \sum_{\mathbf{v}} \Pr[\mathsf{flg}_i(\mathbf{X}, \mathsf{V}) = 0 | X_i \wedge \mathsf{V} = \mathbf{v}] \Pr[X_i \wedge \mathsf{V} = \mathbf{v}]$$

$$= \frac{1}{2}\Pr[\text{ab}_i] + \sum_{\mathbf{v}} \Pr[\text{flg}_i(\mathbf{X}, V) = 0 | V = \mathbf{v}]\Pr[X_i \wedge V = \mathbf{v}] \qquad (7.2.5)$$

$$= \frac{1}{2}\Pr[\text{ab}_i] + \sum_{\mathbf{v}} \Pr[\text{flg}_i(\mathbf{X}, \mathbf{v}) = 0]\Pr[X_i \wedge V = \mathbf{v}]$$

$$= \frac{1}{2}\Pr[\text{ab}_i] + \sum_{\mathbf{v}} \lambda(\mathbf{v})\Pr[X_i \wedge V = \mathbf{v}].$$

Step (7.2.5) follows from Proposition 7.3. So,

$$\Pr[Y_1] - \Pr[Y_2] = \sum_{\mathbf{v}} \lambda(\mathbf{v})(\Pr[X_1 \wedge V = \mathbf{v}] - \Pr[X_2 \wedge V = \mathbf{v}])$$

$$\leq \lambda^+ \sum_{\mathbf{v}} (\Pr[X_1 \wedge V = \mathbf{v}] - \Pr[X_2 \wedge V = \mathbf{v}])$$

$$= \lambda^+ \left(\sum_{\mathbf{v}} \Pr[X_1 \wedge V = \mathbf{v}] - \sum_{\mathbf{v}} \Pr[X_2 \wedge V = \mathbf{v}] \right)$$

$$= \lambda^+ (\Pr[X_1] - \Pr[X_2]).$$

Similarly, $\Pr[Y_1] - \Pr[Y_2] \geq \lambda^-(\Pr[X_1] - \Pr[X_2])$. Combining the two bounds, we get

$$\lambda^-(\Pr[X_1] - \Pr[X_2]) \leq \Pr[Y_1] - \Pr[Y_2] \leq \lambda^+(\Pr[X_1] - \Pr[X_2]). \qquad (7.2.6)$$

Assuming (ε_{dbdh}, t') hardness of DBDH, $|\Pr[Y_1] - \Pr[Y_2]| \leq \varepsilon_{dbdh}$ and so $-\varepsilon_{dbdh} \leq \Pr[Y_1] - \Pr[Y_2] \leq \varepsilon_{dbdh}$. Combining this with (7.2.6), we have

$$-\frac{\varepsilon_{dbdh}}{\lambda^+} \leq \Pr[X_1] - \Pr[X_2] \leq \frac{\varepsilon_{dbdh}}{\lambda^-}. \qquad (7.2.7)$$

From Proposition 7.5, $\frac{1}{\lambda^-} = \frac{1}{\left(1 - \frac{q}{m}\right)\lambda^+}$. Also, for $m > q > 0$, $-\frac{1}{\left(1 - \frac{q}{m}\right)} < -1$. Using these two relations, we obtain

$$-\frac{\varepsilon_{dbdh}}{\left(1 - \frac{q}{m}\right)\lambda^+} \leq \Pr[X_1] - \Pr[X_2] \leq \frac{\varepsilon_{dbdh}}{\left(1 - \frac{q}{m}\right)\lambda^+}.$$

So, $|\Pr[X_1] - \Pr[X_2]| \leq \dfrac{\varepsilon_{dbdh}}{\left(1 - \frac{q}{m}\right)\lambda^+}$. The proof is now completed as follows.

$$\varepsilon_{ibe} = \left| \Pr[X_0] - \frac{1}{2} \right|$$

$$\leq |\Pr[X_0] - \Pr[X_2]|$$

$$\leq |\Pr[X_0] - \Pr[X_1]| + |\Pr[X_1] - \Pr[X_2]|$$

$$\leq \frac{\varepsilon_{dbdh}}{\left(1 - \frac{q}{m}\right)\lambda^+}.$$

□

Putting $m = 4q$ gives us

$$\varepsilon_{ibe} \leq 16nq\varepsilon_{dbdh}.$$

This completes the proof of Theorem 7.1. □

7.3 Generalisation of Waters IBE

Waters IBE provides an elegant solution to the problem of constructing an IBE without the random oracle assumption. However, one disadvantage of the protocol is the requirement of a rather large public parameter. As one can observe from the protocol description, if the identities are n-bit string then one needs an n-length vector $(\overrightarrow{U})$ of elements of G in the public parameter. Independent work by Chatterjee and Sarkar [60] and Naccache [137] provided a generalisation of Waters' IBE. The essential idea is to divide an n-bit identity into l blocks of n/l bits so that the size of the vector $\overrightarrow{U}$ can be reduced from n elements of G to l elements of G. This leads to some associated degradation in the security reduction. In other words, there is a trade-off between the size of the public parameters and the tightness of the security reduction. However, for practical security levels it is possible to ignore the small degradation for appropriate choice of l. Thus the generalisation significantly brings down the public parameter size of Waters IBE.

7.4 Adaptive-Identity Secure HIBE

Waters also proposed a HIBE extending the IBE construction. This gives the first HIBE in the adaptive-identity setting without random oracle. Chatterjee and Sarkar [61] improved upon the Waters HIBE by significantly reducing the public parameter size. This construction is outlined below.

For an h-HIBE the identities are of the type $(\mathrm{id}_1,\ldots,\mathrm{id}_j)$, where $j \leq h$ and each $\mathrm{id}_k = (\mathrm{id}_{k,1},\ldots,\mathrm{id}_{k,l})$, where $\mathrm{id}_{k,i}$ is an (n/l)-bit string (assuming l divides n). Choosing $l = n$ gives id_k to be an n-bit string as considered by Waters.

Set-Up: The protocol is described in the symmetric pairing setting (p,G,G,G_T,e). The public parameters are the following elements: P, $P_1 = xP$, P_2, $U_1',\ldots,U_h'$, $U_1,\ldots,U_l$, where $G = \langle P \rangle$, x is chosen randomly from $\mathbb{Z}_p$ and the other quantities are chosen randomly from G.

The master secret is xP_2. The quantities P_1 and P_2 are not directly required; instead $e(P_1,P_2)$ is required. Hence one may store $e(P_1,P_2)$ as part of the public parameters instead of P_1 and P_2. (In fact, this optimization is possible for all (H)IBE schemes that fall under the Boneh-Boyen paradigm.)

A Useful Shorthand:

Let $v = (v_1, \ldots, v_l)$, where each v_i is an (n/l)-bit string and is considered to be an element of $\mathbb{Z}_{2^{n/l}}$. For $1 \leq k \leq h$ define,

$$V_k(v) = U'_k + \sum_{i=1}^{l} v_i U_i. \qquad (7.4.8)$$

When v is clear from the context we will write V_k instead of $V_k(v)$. The modularity introduced by this notation allows an easier understanding of the protocol.

Key-Gen: Let $\mathsf{id} = (\mathsf{id}_1, \ldots, \mathsf{id}_j)$, $j \leq h$, be the identity for which the private key is required. Choose $r_1, \ldots, r_j$ randomly from $\mathbb{Z}_p$ and define $d_{\mathsf{id}} = (d_0, d_1, \ldots, d_j)$ where $d_0 = xP_2 + \sum_{k=1}^{j} r_k V_k(\mathsf{id}_k)$ and $d_k = r_k P$ for $1 \leq k \leq j$.

Key delegation can be done in the usual way as shown in the construction of BB-HIBE in Chapter 5. Suppose $(d'_0, d'_1, \ldots, d'_{j-1})$ is a private key for the identity $(\mathsf{id}_1, \ldots, \mathsf{id}_{j-1})$. To generate a private key for id, first choose r_j randomly from $\mathbb{Z}_p$ and compute d_{id} as follows: $d_0 = d'_0 + r_j V_j(\mathsf{id}_j)$; $d_i = d'_i$ for $1 \leq i \leq j-1$; $d_j = r_j P$.

Encrypt: Let $\mathsf{id} = (\mathsf{id}_1, \ldots, \mathsf{id}_j)$ be the identity under which a message $M \in G_T$ is to be encrypted. Return

$$(C_0 = M \times e(P_1, P_2)^t, C_1 = tP, B_1 = tV_1(\mathsf{id}_1), \ldots, B_j = tV_j(\mathsf{id}_j)),$$

where t is a random element of $\mathbb{Z}_p$.

Decrypt: Let $C = (C_0, C_1, B_1, \ldots, B_j)$ be a ciphertext encrypted for the identity $\mathsf{id} = (\mathsf{id}_1, \ldots, \mathsf{id}_j)$ and $(d_0, d_1, \ldots, d_j)$ is the corresponding decryption key. Return

$$C_0 \times \frac{\prod_{k=1}^{j} e(B_k, d_k)}{e(d_0, C_1)}.$$

It is easy to verify the correctness of decryption.

Note that for the jth level of the HIBE, the protocol only adds a single element, i.e., U'_j in the public parameter while the elements $U_1, \ldots, U_l$ are re-used for each level. This way it is possible to shorten the public parameter size. The novelty is that this reduction in the public parameter size is achieved without any associated degradation in the security reduction. Also note that, if $h = 1$, then we obtain an IBE scheme and if further $l = n$, then we obtain Waters' IBE.

7.5 Converting to a CCA-Secure HIBE

The HIBE scheme of Section 7.4 can be modified to achieve CCA-security. This modification is based on the technique used by Boyen, Mei and Waters [49] as explained in Chapter 6. Symmetric key techniques are combined with the CPA-secure HIBE to obtain a hybrid construction. Several pairing computations can be eliminated by using symmetric key authenticated encryption (AE) to check for ciphertext validity. The details of the CCA-secure scheme are given in Figure 7.1. Bold entries denote the portions that are introduced to obtain CCA-security over and above the CPA-secure scheme in Section 7.4.

The scheme uses three extra primitives which have not been considered so far. Below we provide a brief description of these primitives and refer the reader to places where more details may be found.

Authenticated Encryption. This is a symmetric key primitive which combines the dual role of encryption and authentication into a single functionality. There are two algorithms associated with the primitive – AE.Encrypt and AE.Decrypt. The algorithm AE.Encrypt takes as input a nonce (or initialization vector IV) and the message M and uses the secret key dk to produce the ciphertext (cpr, tag). Typically, the length of cpr is equal to that of M and tag is some extra information which provides authentication. Algorithm AE.Decrypt takes (cpr, tag) as input and provides as output either the corresponding message M or the symbol $\perp$ which denotes that the authentication has failed. Intuitively, an AE scheme provides confidentiality of the message and at the same time ensures that active tampering of the ciphertext by an adversary will be detected. Formal description of AE schemes was first independently proposed in [22, 119]. It is not too difficult to design AE schemes using block ciphers which make two passes over the data. Somewhat counter-intuitively, several one-pass AE schemes [114, 96, 149] have been constructed the most famous among which is the OCB mode [149]. See [57, 153] for a general family of one-pass AE schemes as well as for schemes which offer certain advantages over OCB.

Universal One-Way Hash Family (UOWHF). This is a family of hash functions $\{H_s\}_{s \in \mathscr{S}}$ where each H_s has the same domain and range. An adversarial game for the family is defined as follows. The adversary chooses x from the domain; is then given a uniform random s and has to find an x' distinct from x such that $H_s(x) = H_s(x')$. Compared to the usual notion of collision resistance, this game is more difficult for the adversary since the adversary has to commit to one of the inputs even before knowing the function for which a collision has to be found. Viewed differently, the requirement on the hash function is lesser and so it may be more desirable to base a scheme on a UOWHF rather than on a collision-resistant hash function. The notion of UOWHF was introduced by Naor and Yung [140] and was analysed in the concrete security setting by Bellare and Rogaway [25]. For later practically oriented work on extending the domain of a UOWHF see [159, 152].

Key Derivation Function (KDF). A KDF function maps a domain to a range in a manner such that if the input to the KDF is a random element of the domain, then the output of the KDF is indistinguishable from random to a computationally bounded adversary. This notion was introduced by Shoup [162] as a component in constructing hybrid PKE schemes.

The formal security reduction is long. We provide some idea of the argument with reference to Figure 7.1. The scheme is obtained by modifying the previous CPA-secure scheme. So, the technique for simulating key extraction queries is already built into the system. Also, for the challenge ciphertext, all elements except for C_2 can be properly generated. The additional mechanism is used to ensure decryption queries can be properly handled and that the component C_2 of the challenge ciphertext can be properly generated.

Essentially, there are two separate encapsulations of the session key K. The first is using the CPA-secure HIBE scheme and the second is using the Boneh-Boyen selective-identity secure IBE scheme. The identity for the first encapsulation is the actual identity for which encryption is done while the "identity" for the second encapsulation is the output of the function H_s. This is the basic idea of Boyen-Mei-Waters transformation of a CPA-secure IBE to a CCA-secure PKE discussed in Chapter 6.

Note that the length j of the identity tuple is part of the input to H_s. This binding of the length of the identity tuple ensures that given the challenge ciphertext, it is not possible to trivially make a decryption query on a proper prefix of the challenge identity by simply discarding some elements from the challenge ciphertext.

If a decryption query is made for an identity which is not equal to the challenge identity or one of its prefixes, then (as is standard) the technique of simulating key extraction queries is used to generate a proper decryption key for this identity and use that to decrypt the ciphertext and answer the query. The problem arises when the identity is either the challenge identity or one of its prefixes. In this case, it can be argued that the inputs to H_s for the decryption query and for the challenge ciphertext are necessarily different. So, (using the UOWHF property) the outputs are also different. Since this output is an "identity" for the second encapsulation through the in-built BB-IBE scheme, the technique for simulating key extraction queries in the BB-IBE scheme can be used to obtain a decryption key for this identity and then use it to decrypt the ciphertext. In a similar manner it is possible to show that the element C_2 can be properly generated.

Suppose that the adversary makes a mal-formed decryption query where the elements

$$C_1, C_2, B_1, \ldots, B_j$$

do not satisfy the relation to each other as required, i.e., they are not formed by using the same randomiser. The relation between C_1 and C_2 is explicitly checked by the pairing computation in the decryption algorithm. But, the relation between C_1 and $B_1, \ldots, B_j$ is not explicitly checked. If the required relation (i.e., the discrete log to the respective bases is t) does not hold, then it can be argued that the session key K generated in the decryption algorithm is a random element which is independent of the K^* implicitly defined by the challenge ciphertext. The decryption algorithm of the AE scheme will be invoked with a random key K and for which the adversary has not earlier seen any ciphertext. So, the authentication mechanism of the AE scheme will generate an invalid ciphertext error and $\perp$ will be returned to the adversary. Note that the crucial issue here is that the relation between C_1 and $B_1, \ldots, B_j$ is implicitly verified using the AE scheme. Doing this directly would have required a number of pairing computations which would make the decryption algorithm less efficient. The pairing based approach for verifying well-formedness of the ciphertext has been used in [120].

Fig. 7.1 CCA-secure HIBE.
1. Maximum depth of the HIBE is h.
2. Identities are of the form $id = (id_1, \ldots, id_j)$, $j \in \{1, \ldots, h\}$, $id_k = (id_{k,1}, \ldots, id_{k,l})$ and $id_{k,i}$ is an (n/l)-bit string.
3. The setting $(p, G = \langle P \rangle, G, G_T, e)$ is of Type 1 pairing.
4. The notation $V_k()$ is given in (7.4.8).
5. It is possible to avoid computing the pairing value $e(P_1, P_2)$ during encryption by replacing P_2 in the public parameters by $e(P_1, P_2)$.

HIBE.Set-Up	
1. Choose α randomly from $\mathbb{Z}_p$.	
2. Set $P_1 = xP$.	HIBE.KeyGen: Identity $id = (id_1, \ldots, id_j)$.
3. Choose $P_2, U_1', \ldots, U_h', U_1, \ldots, U_l$ randomly from G.	
	1. Choose $r_1, \ldots, r_j$ randomly from $\mathbb{Z}_p$.
4. Choose W randomly from G.	2. $d_0 = xP_2 + \sum_{k=1}^{j} r_k V_k(id_k)$.
5. Let $\mathbf{H_s} : \{1, \ldots, h\} \times \mathbf{G} \rightarrow \mathbb{Z}_p$ be chosen from a UOWHF and made public.	3. $d_k = r_k P$ for $k = 1, \ldots, j$.
	4. Output $d_{id} = (d_0, d_1, \ldots, d_j)$.
6. Public parameters: $P, P_1, P_2, U_1', \ldots, U_h', U_1, \ldots, U_l$ and **W**.	
7. Master secret key: xP_2.	
	HIBE.Decrypt:
HIBE.Encrypt:	Identity $id = (id_1, \ldots, id_j)$;
Identity $id = (id_1, \ldots, id_j)$; message M.	ciphertext $(C_1, \mathbf{C_2}, B_1, \ldots, B_j, cpr, tag)$; decryption key $d_{id} = (d_0, d_1, \ldots, d_j)$.
1. Choose t randomly from $\mathbb{Z}_p$.	
2. $C_1 = tP$, $B_1 = tV_1(id_1), \ldots, B_j = tV_j(id_j)$.	1. $\gamma = \mathbf{H_s(j, C_1)}$; $\mathbf{W}_\gamma = \mathbf{W} + \gamma\mathbf{P_1}$.
3. $K = e(P_1, P_2)^t$.	2. **If $e(\mathbf{C_1}, \mathbf{W}_\gamma) \neq e(\mathbf{P}, \mathbf{C_2})$ return $\perp$.**
4. $(IV, dk) = KDF(K)$.	3. $K = e(d_0, C_1) \times \prod_{k=1}^{j} e(B_k, -d_k)$.
5. $(cpr, tag) = AE.Encrypt_{dk}(IV, M)$.	4. $(IV, dk) = KDF(K)$.
6. $\gamma = \mathbf{H_s(j, C_1)}$; $\mathbf{W}_\gamma = \mathbf{W} + \gamma\mathbf{P_1}$; $\mathbf{C_2} = t\mathbf{W}_\gamma$.	5. $M = AE.Decrypt_{dk}(IV, cpr, tag)$. (This may abort and return $\perp$).
7. Output $(C_1, \mathbf{C_2}, B_1, \ldots, B_j, cpr, tag)$.	6. Output M.

7.6 Further Applications of Waters Technique

The map $(id_1, \ldots, id_n) \mapsto id_1 U_1 + \cdots + id_n U_n$ first used by Waters [169] for designing an IBE scheme can be used for obtaining other IBE schemes. Boneh, Boyen and Goh [35] proposed a selective-identity secure constant size ciphertext HIBE. This construction has been discussed in Chapter 5. It had been suggested in this work that the hashing technique used by Waters in [169] can be used to obtain an adaptive-identity secure constant size ciphertext HIBE. The details have been worked out in a separate paper [62]. Setting the number of levels in the HIBE to one provides another IBE scheme which is secure against adaptive-identity attacks.

In a related work, Kiltz and Vahlis [121], present yet another adaptive-identity secure IBE scheme. Security is based on a different problem where an instance provides extra information compared to the DBDH problem. CCA-security is attained by using symmetric key techniques to verify the well-formedness of the cipher-text. This is reminiscent of similar techniques used in the context of PKE schemes by Kurosawa and Desmedt [127] and first used in the context of IBE schemes by Sarkar and Chatterjee [154].

7.7 Conclusion

In this chapter we have described several important (H)IBE schemes. Waters IBE protocol and its security are analyzed in depth. Waters constructions and their later generalisations and improvements resolved the important problem of obtaining a practical (H)IBE in the standard model. However, the problem of constructing an IBE and HIBE in the standard model with a tight reduction still remained open. In the next chapter we will discuss how the problem of obtaining a tighter reduction for IBE (and HIBE) is addressed.

Chapter 8
Further IBE Constructions

The Waters 2005 IBE scheme and its extensions provide practical solutions to the problem of constructing IBE schemes which are secure against adaptive-identity attacks based on a well-accepted complexity assumption (the DBDH problem) and do not use random oracles in the security reduction. Some questions are still of interest.

1. These IBE schemes suffer from a security degradation by a factor of q which is the number of key extraction queries. Is it possible to obtain a tight security reduction to the underlying hard problem?
2. The size of public parameters in Waters 2005 IBE scheme is rather large. The extensions by Chatterjee-Sarkar and Naccache reduce the size. But, still asymptotically speaking, the size of the public parameters grows with an increase in the value of the security parameter. Is it possible to delink the size of the public parameters from the value of the security parameter?

Answers to these two questions have been proposed in the literature and in this chapter we discuss the corresponding schemes.

Gentry [91] proposed a construction of an IBE with a tight security reduction based on a hard computational problem. This may be taken as a positive answer to the first question above. But, the answer is not complete. The hard problem used by Gentry is *not* a standard problem in the pairing settings. In fact, it is a non-static problem, i.e., the size of a problem instance depends on the maximum number of key-extraction queries that the adversary is allowed to make. It is not clear whether obtaining a tight reduction based on such a problem actually promises better security than obtaining a looser reduction based on a more simple problem such as the DBDH problem.

The second question was addressed by Waters in 2009 [170]. He proposed a construction where the number of group elements required in the public parameters is independent of the value of the security parameter. This provides a positive answer to the second question. The underlying hard problems are the DBDH and the decision linear problem. However, the security reduction is not tight and the degradation is by a factor of q as in the case of Waters 2005 IBE scheme. On the other hand,

when extended to HIBE, Waters technique yields a construction where it is possible
to prevent the exponential degradation in the security reduction with the number of
levels.

8.1 Gentry's IBE

Broadly speaking the strategy followed in most of the IBE schemes prior to the
work of Gentry is to implicitly divide the set of identities into two disjoint subsets.
For one subset, the simulator is able to generate decryption keys while for the other
subset, the simulator is able to generate challenge ciphertexts. This is the so-called
partitioning approach. If the adversary asks for a private key of an identity for which
it is not possible to generate the key, the simulator has to abort. Similarly, if the
adversary provides a challenge identity for which the simulator is unable to generate
a ciphertext, even then the simulator has to abort. The probabilities of these abort
conditions taken together lead to a degradation of the security bound.

The main contribution of Gentry's work is to obtain a tight security reduction. We
try to give an intuitive idea of how this is achieved. Consider the problem of obtaining a CPA-secure IBE scheme. The adversary can make key-extraction queries and
ask for the challenge ciphertext. (Since we are considering CPA-security, the adversary is not allowed to make decryption queries.) So, the simulator in the proof has
to answer the key-extraction queries and also generate a ciphertext for the challenge
identity. The goal of the simulator is to convert the adversary's ability to distinguish
between encryptions of the two messages into an algorithm to solve an instance of
the underlying hard problem. For this, the simulator expects the adversary to be able
to do some task which it cannot do itself. Suppose, for example, that the simulator
is able to generate decryption key for the challenge identity. Then it could itself
decrypt the challenge ciphertext. So, working with the adversary does not provide
the simulator with any useful information. It appears that the partitioning strategy
mentioned above might be an inherent one, i.e., the simulator should not be able to
generate keys for the challenge identity.

Gentry overcomes this apparent problem using a crucial observation. Since the
key extraction algorithm is probabilistic, for each identity there is a large set of
possible keys. The key extraction algorithm can be seen to be sampling from this
set. Now, suppose that for each identity, the simulator can generate exactly one decryption key (even a *small* set would do). So, even for the challenge identity, the
simulator is able to generate exactly one decryption key. The adversary, however,
does not know which key (out of the possible large set of keys) the simulator can
generate. So, in its attack on the IBE scheme, with very high probability, the adversary is likely to use a different key. Then the simulator obtains some non-trivial
information from the adversary which it can utilize to attack the underlying hard
problem.

Gentry divides the presentation of his IBE scheme into two parts. The first
part describes a CPA-secure scheme and the second part describes a CCA-secure

scheme. We follow his presentation. In Section 8.1.1, we describe the CPA-secure scheme and the security reduction. The CCA-secure scheme is described in Section 8.1.2 but we do not provide the details of the security reduction.

8.1.1 CPA-Secure Construction

Before going into the construction of the scheme we mention two things.

First, the scheme actually provides recipient anonymity, i.e., it is an anonymous-IBE. This notion has been described in Chapter 2 and we briefly recall the concept. The idea is that given a ciphertext it is computationally infeasible for an adversary to determine for which identity it was generated. Formally, the notion of anonymity is captured by asking the adversary in the challenge stage, to provide a pair of identities $(\mathsf{id}_0, \mathsf{id}_1)$ along with a pair of equal length messages (M_0, M_1). A pair of independent and uniform random bits (b, c) is chosen and the challenge ciphertext is generated for the message M_b under the identity id_c which is then provided to the adversary. In this model, the adversary has to guess both the bits b and c with probability significantly away from $1/4$.

Second, the security of the scheme is based on the hardness of the (truncated) decision q-ABDHE problem (see Chapter 3). An instance of this problem consists of a tuple $(P', P'_{q+2}, P, P_1, \ldots, P_q, Z)$ where $P_i = \alpha^i P$ for some random $\alpha \in \mathbb{Z}_p$. The task is to decide whether Z is equal to $e(P_{q+1}, P')$ or Z is a random element of G_T. Here q is a parameter of the problem instance. In the proof relating the security of the IBE scheme to this problem, q will depend on the maximum number of key-extraction queries the adversary is allowed to make. In other words, the hardness assumption required for security is a non-static one.

We now describe the CPA-secure scheme of Gentry followed by its security reduction.

Set-Up: The PKG chooses independent and uniform random elements P and Q from G; a uniform random element α from $\mathbb{Z}_p$; and sets $P_1 = \alpha P$. The public parameters consist of (P, P_1, Q) while the master secret key is α. Identities are elements of $\mathbb{Z}_p$.

Key-Gen: The input is an identity $\mathsf{id} \in \mathbb{Z}_p$ and the PKG generates a decryption key d_{id} for this identity. To do this, the PKG chooses a uniform random $r \in \mathbb{Z}_p$ and sets $d_{\mathsf{id}} = (r, Q_{\mathsf{id}})$, where

$$Q_{\mathsf{id}} = \frac{1}{\alpha - \mathsf{id}} (Q - rP).$$

If $\mathsf{id} = \alpha$, then the PKG aborts; an event which occurs with probability $1/p$.

Encrypt: The input consists of a message $M \in G_T$ and an identity $\mathsf{id} \in \mathbb{Z}_p$. The sender generates a uniform random $t \in \mathbb{Z}_p$ and computes the ciphertext to be (U, V, W) where

$$U = t(P_1 - \text{id}P); \quad V = e(P,P)^t; \quad W = M \times e(P,Q)^{-t}.$$

Decrypt: On input (U,V,W); identity id; and decryption key $d_{\text{id}} = (r,Q_{\text{id}})$, the receiver outputs

$$W \times e(U,Q_{\text{id}}) \times V^r.$$

If the ciphertext and the decryption key are proper, then the correctness of the decryption can be seen by the following computation.

$$
\begin{aligned}
e(U,Q_{\text{id}}) \times V^r &= e(t(\alpha - \text{id})P, \frac{1}{\alpha - \text{id}}(Q - rP)) \times e(P,P)^{tr} \\
&= e(tP, Q - rP) \times e(P,P)^{tr} \\
&= e(P,Q)^t \times e(P,P)^{-tr} \times e(P,P)^{tr} \\
&= e(P,Q)^t.
\end{aligned}
$$

The security statement for this scheme is given below.

Theorem 8.1. *let* $q = q_{\text{id}} + 1$, *where* q_{id} *is the number of key extraction queries. The above scheme is* $(t,\varepsilon,q_{\text{id}})$ ANON-IND-ID-CPA *secure if the truncated decision* (t',ε',q)-ABDHE *assumption holds, where* $\varepsilon = \varepsilon' + 2/p$ *and* $t' = t + O(q^2\tau)$, τ *is the time for a scalar multiplication in G.*

Proof : Let $\mathscr{A}$ be a $(t,\varepsilon,q_{\text{id}})$-adversary for the ANON-IND-ID-CPA game. This is used to construct an algorithm $\mathscr{B}$ which solves the truncated decision q-ABDHE problem. $\mathscr{B}$ takes as input a tuple $(P', P'_{q+2}, P, P_1, \ldots, P_q, Z)$, where Z is either $e(P_{q+1}, P')$ or a random element of G_T. Note that $P_i = \alpha^i P$, for some random $\alpha \in \mathbb{Z}_p$. Algorithm $\mathscr{B}$ proceeds as follows.

Set-Up: $\mathscr{B}$ generates a random polynomial $f(x)$ of degree q with coefficients in $\mathbb{Z}_p$ and sets $Q = f(\alpha)P$. Note that Q can be computed by using the elements $P, P_1, \ldots, P_q$. The public key (P, P_1, Q) is given to the adversary. Since P, α and $f(x)$ are chosen independently and uniformly at random, the distribution of the public key is the same as that in the actual scheme.

Key Extraction Query: This can happen in both Phase 1 and Phase 2 and are tackled in the same manner in both phases. Suppose $\mathscr{A}$ asks for the decryption key of an identity id $\in \mathbb{Z}_p$. If id $= \alpha$, then $\mathscr{B}$ can immediately solve the given instance of truncated q-ABDHE problem. Otherwise, let $F_{\text{id}}(x)$ be the degree $(q - 1)$ polynomial $(f(x) - f(\text{id}))/(x - \text{id})$. $\mathscr{B}$ now defines the private key for id to be $(f(\text{id}), F_{\text{id}}(\alpha)P)$. The fact that this is a valid private key for id can be seen from the following simple computation.

$$F_{\text{id}}(\alpha)P = \frac{f(\alpha) - f(\text{id})}{\alpha - \text{id}}P = \frac{1}{\alpha - \text{id}}(f(\alpha)P - f(\text{id})P) = \frac{1}{\alpha - \text{id}}(Q - f(\text{id})P).$$

Challenge: Suppose $\mathscr{A}$ outputs two identities id_0, id_1 and two messages M_0, M_1. As in the case of simulation of key extraction queries, if α is either id_0 or id_1, then $\mathscr{B}$ can immediately solve the given instance of the truncated q-ABDHE problem. So, assume that this is not the case. $\mathcal{B}$ generates two independent and uniform random bits b, c and computes a private key (r, Q_{id_b}) for id_b using the method for simulating key extraction queries.

Let $f'(x) = x^{q+2}$ and let

$$
\begin{aligned}
F'_{\mathrm{id}_b}(x) &= \frac{f'(x) - f'(\mathrm{id}_b)}{(x - \mathrm{id}_b)} \\
&= \frac{x^{q+2} - \mathrm{id}_b^{q+2}}{(x - \mathrm{id}_b)} \\
&= x^{q+1} + \mathrm{id}_b x^q + \cdots + \mathrm{id}_b^q x + \mathrm{id}_b^{q+1}.
\end{aligned}
$$

This is a polynomial of degree $(q+1)$. Let $P_0 = P$. The challenge ciphertext consists of (U, V, W) where

$$
U = (f'(\alpha) - f'(\mathrm{id}_b))P';
$$

$$
V = Z \times e\left(P', \sum_{i=0}^{q} F'_{\mathrm{id}_b, i} P_i\right);
$$

$$
W = M_c / (e(U, Q_{\mathrm{id}_b}) \times V^r).
$$

In the above, $F'_{\mathrm{id}_b, i} = \mathrm{id}_b^{q+1-i}$ is the coefficient of x^i in the polynomial $F'_{\mathrm{id}_b}(x)$. In effect, the sum $\sum_{i=0}^{q} F'_{\mathrm{id}_b, i} P_i$ is equal to

$$
\begin{aligned}
F'_{\mathrm{id}_b}(\alpha)P - \alpha^{q+1}P &= (\mathrm{id}_b x^q + \cdots + \mathrm{id}_b^q x + \mathrm{id}_b^{q+1})P \\
&= \mathrm{id}_b P_q + \cdots + \mathrm{id}_b^q P_1 + \mathrm{id}_b^{q+1} P_0.
\end{aligned}
$$

This can be computed from the knowledge of the coefficients $F'_{\mathrm{id}_b, i}$s and the P_is and without the knowledge of α.

Guess: At the end, $\mathcal{A}$ outputs guesses b' and c' of b and c respectively. If $b = b'$ and $c = c'$, then $\mathcal{B}$ outputs 1 (indicating that Z is real), otherwise it outputs 0.

As already noted, the distribution of the public parameters is as required by the actual scheme. Further, if Z is real, i.e., $Z = e(P_{q+1}, P')$, then the distribution of the ciphertext is also the same as in the actual scheme. This can be seen from the following computation. Note that $f'(\alpha) - f'(\mathrm{id}_b) = (\alpha - \mathrm{id}_b)F'_{\mathrm{id}_b}(\alpha)$. Let $s = (\log_P P')F'_{\mathrm{id}_b}(\alpha)$ and then $U = s(\alpha - \mathrm{id}_b)P$. Now, using the value of Z

$$
\begin{aligned}
V &= e(P_{q+1}, P') \times e\left(P', \sum_{i=0}^{q} F'_{\mathrm{id}_b, i} P_i\right) \\
&= e(P', F'_{\mathrm{id}_b}(\alpha)P) \\
&= e(P, P)^s
\end{aligned}
$$

and so

$$\frac{M_c}{W} = e(U, Q_{id_b}) \times V^r = e(P, Q)^s.$$

This follows because U and V are of proper form and (r, Q_{id_b}) is a proper private key for id_b. The details of this calculation is similar to the one given to show the correctness of decryption.

Consider a set $\mathscr{I}$ consisting of α, id_b and the identities queried by $\mathscr{A}$. The independent and uniform random distribution of the decryption keys will follow if it holds that the values $f(a)$ ($a \in \mathscr{I}$) are independent and uniform random. This follows from the observations that $|\mathscr{I}| \leq q + 1$ and f is a polynomial of degree q whose coefficients are chosen independently and uniformly at random from $\mathbb{Z}_p$.

If Z is real, then the simulation is perfect and let X_0 be the event that $\mathscr{A}$'s guesses are both correct. Let X_1 be the event that the adversary's guesses are correct when Z is a uniform random element of G_T. Clearly,

$$\Pr[X_0] - \Pr[X_1] = \Pr[\mathscr{B} \text{ outputs } 1 | Z \text{ is real}] - \Pr[\mathscr{B} \text{ outputs } 1 | Z \text{ is random}] \leq \varepsilon'.$$

Now we argue that when Z is random, the ciphertext statistically hides the bits b and c from the adversary. The challenge ciphertext consists of three elements U, V and W. We show that the randomness of these three elements are determined by three independent and uniform random quantities α, Z and r which are themselves independent of b and c. Here r is part of the private key and is computed as $r = f(id_b)$. Since f is a polynomial with coefficients chosen uniformly at random from $\mathbb{Z}_p$, r is uniformly distributed.

The independent and uniform random choices of α and Z ensure that U and V are independent and uniform random. Let E be the event that $V = e(U, P)^{1/(\alpha - id_0)}$ or $V = e(U, P)^{1/(\alpha - id_1)}$. Then $\Pr[E] \leq 2/p$. If E does not occur, then by the uniform random choice of r

$$e(U, Q_{id_b}) \times V^r = e\left(U, \frac{1}{\alpha - id_b}(Q - rP)\right) \times V^r$$

$$= e(U, Q)^{1/(\alpha - id_b)} \times \left(\frac{V}{e(U, P)^{1/(\alpha - id_b)}}\right)^r$$

is a uniform random element of G_T and so U, V and W are independent and uniform random elements. So, these perfectly hide the bits b and c from $\mathscr{A}$. It follows that $\Pr[X_1 | \overline{E}] = 1/4$. Further,

$$\Pr[X_1] = \Pr[X_1 | E] \Pr[E] + \Pr[X_1 | \overline{E}] \Pr[\overline{E}] \leq \Pr[E] + \Pr[X_1 | \overline{E}]$$

and so $\Pr[X_1] - \Pr[X_1 | \overline{E}] \leq \Pr[E]$.

If we let $\mathscr{A}$ to be an adversary which maximizes the advantage of breaking the IBE scheme among all adversaries that run in time t and make q_{id} key extraction queries, then we have

$$\varepsilon = \Pr[X_0] - \frac{1}{4} = \Pr[X_0] - \Pr[X_1 | \overline{E}]$$

$$= \Pr[X_0] - \Pr[X_1] + \Pr[X_1] - \Pr[X_1 | \overline{E}]$$
$$\leq \varepsilon' + \frac{2}{p}.$$

The time for $\mathscr{B}$ to simulate the queries of $\mathscr{A}$ is dominated by the time to compute $F_{id}(\alpha)P$ where $F_{id}(x)$ is a polynomial of degree $q-1$. Each evaluation requires $O(q)$ scalar multiplications in G for a total of $O(q^2)$ scalar multiplications overall. $\square$

8.1.2 CCA-Secure Construction

In Chapter 6 we have seen several techniques for converting a CPA-secure IBE scheme to a CCA-secure scheme. Both the generic and the non-generic techniques use a 2 level CPA-secure HIBE to construct a CCA-secure IBE. However, the CCA-secure scheme given by Gentry [91] does not use any of these techniques. Instead, Gentry adapts techniques used in the construction of the PKE scheme by Cramer and Shoup [74] to the current setting. In the security reduction of the Cramer-Shoup PKE scheme, the simulator knows exactly one valid decryption key. This combines very well with the CPA-secure IBE scheme described in the previous section, where the simulator can generate exactly one valid decryption key for each identity. The ability to generate decryption keys for any identity implies that the simulator is able to answer any decryption query. Cramer and Shoup show in their simulation that certain quantities used to obtain the decryption key remain statistically hidden from the adversary. The technical novelty in Gentry's proof is to modify this strategy to work in the identity-based setting.

Below we provide the description of the CCA-secure IBE scheme. Details of the security reduction are not provided.

Setup: The PKG chooses independent and uniform random elements P and Q_1, Q_2, Q_3 from G; a uniform random element α from $\mathbb{Z}_p$; and sets $P_1 = \alpha P$. It also chooses a function $H : G \times G_T^2 \to \mathbb{Z}_p$ from a family of universal one-way hash functions. The public parameters consist of (P, P_1, Q_1, Q_2, Q_3) and H while the master secret key is α. Identities are elements of $\mathbb{Z}_p$.

Key-Gen: The input is an identity $id \in \mathbb{Z}_p$ and the PKG generates a decryption key d_{id} for this identity. To do this, the PKG chooses independent and uniform random r_1, r_2 and $r_3 \in \mathbb{Z}_p$ and sets $d_{id} = (r_1, r_2, r_3, Q_{id,1}, Q_{id,2}, Q_{id,3})$, where

$$Q_{id,i} = \frac{1}{\alpha - id}(Q_i - r_i P).$$

If $id = \alpha$, then the PKG aborts – an event which occurs with probability $1/p$.

Encrypt: The input consists of a message $M \in G_T$ and an identity $id \in \mathbb{Z}_p$. The sender generates a uniform random $t \in \mathbb{Z}_p$ and computes the ciphertext to be (U, V, W, X) where

$$U = t(P_1 - \text{id}P); \quad V = e(P,P)^t; \quad W = M \times e(P,Q_1)^{-t};$$
$$\beta = H(U,V,W); \quad X = e(P,Q_2)^t \times e(P,Q_3)^{t\beta}.$$

Decrypt: The input consists of ciphertext (U,V,W,X), identity id and decryption key $d_{\text{id}} = (r_1, r_2, r_3, Q_{\text{id},1}, Q_{\text{id},2}, Q_{\text{id},3})$. The receiver performs the following steps.

1. Compute $\beta = H(U,V,W)$;
2. verify whether $X = e(U, Q_{\text{id},2} + \beta Q_{\text{id},3}) \times V^{(r_2 + \beta r_3)}$;
   if the test fails, then the output is $\perp$;
3. return $W \times e(U, Q_{\text{id},1}) \times V^{r_1}$.

If the ciphertext and the decryption key are proper, then the ciphertext passes the test as seen by the following calculation.

$$e(U, Q_{\text{id},2} + \beta Q_{\text{id},3}) \times V^{(r_2 + \beta r_3)}$$
$$= e\left(t(\alpha - \text{id})P, \frac{1}{\alpha - \text{id}}(Q_2 + \beta Q_3) - \frac{r_2 + r_3\beta}{\alpha - \text{id}}P\right) \times e(P,P)^{t(r_2 + \beta r_3)}$$
$$= e\left(t(\alpha - \text{id})P, \frac{1}{\alpha - \text{id}}(Q_2 + \beta Q_3)\right)$$
$$= e(P, Q_2)^t \times e(P, Q_3)^{t\beta}.$$

Correctness of the message recovery is shown by the following calculation.

$$e(U, Q_{\text{id},1}) \times V^{r_1} = e(t(\alpha - \text{id})P, \frac{1}{\alpha - \text{id}}(Q_1 - r_1P)) \times e(P,P)^{tr_1} = e(P, Q_1)^t.$$

8.1.3 Previous and Further Work

The security of Gentry's scheme is based on the assumption that the truncated decision q-ABDHE problem is hard. This is a non-static hardness assumption where the size of an instance depends on the number of queries that an adversary is allowed to make. (Recall that the other kind of non-static assumption is where the size of an instance depends on the depth of a HIBE.) Gentry's work is not the first to use such an assumption. Previously the bilinear Diffie-Hellman inversion (BDHI) assumption had been used to give a security reduction of Sakai and Kasahara IBE [151] and also by Boneh and Boyen [32]. An attempt [46] has been made to explain these two works in a general framework called the *ad-hoc* encryption from exponent inversion assumption. This framework, however, does not capture Gentry's construction. Perhaps one reason for this is that Gentry's construction has the best features among all the constructions that use an assumption where the instance size depends on the number of queries.

Kiltz and Vahlis [121] provide a different modification of the CPA-secure scheme of Gentry to attain CCA-security. The purpose of the modification is to improve the efficiency of the scheme. This efficiency improvement is achieved by using symmetric key authentication techniques to check for ciphertext well-formedness. In the context of PKE this idea was first used by Kurosawa and Desmedt [127] while in the context of IBE this was first used by Sarkar and Chatterjee [154].

In a subsequent work, Gentry and Halevi [92] extended the idea behind Gentry's IBE scheme to the hierarchical setting. They required a more complicated assumption and the construction itself is also quite complicated. Importance of this work rests on the fact that it is the first to provide a HIBE scheme whose security does not degrade exponentially with the number of levels. Later it was shown by Waters [170] that the same result can be achieved using a much simpler method.

8.2 Dual System Encryption

In the introductory part of this chapter we have noted that Waters [170] recently proposed a new paradigm called dual system encryption. He proposed the construction of an IBE scheme in this paradigm with some novel features. We will call this scheme the Waters-2009 IBE to distinguish it from the earlier IBE scheme also by Waters [169] described in Chapter 7.

The main difficulty in the security reduction of an IBE scheme is the simulation of key extraction queries. Earlier we have seen two main strategies for performing such simulation. In the partitioning strategy, the simulator implicitly partitions the identity space into two parts and is able to answer key extraction queries for identities from one part and generate challenge ciphertext for identities from the other part. This approach leads to a security degradation because the simulator has to abort the game on certain queries. The previous section describes an approach by Gentry on how to avoid using the partitioning strategy and hence obtain a tight reduction. This is based on the fact that the simulator is able to generate exactly one key for every identity.

Waters-2009 IBE scheme uses a somewhat different approach in the security reduction. Apart from usual ciphertexts and usual decryption keys, he defines two new notions – semi-functional ciphertexts and semi-functional decryption keys. These have some special properties.

1. A semi-functional ciphertext *can* be decrypted using a normal decryption key.
2. A normal ciphertext *can* be decrypted using a semi-functional decryption key.
3. A semi-functional ciphertext *cannot* be decrypted using a semi-functional decryption key.

Waters proposed a hybrid argument where the reduction proceeds through a sequence of $(q+3)$ games where q is the maximum number of key extraction queries. The first game is the usual security game defining the CPA-security of IBE. In the second game, the normal challenge ciphertext is replaced by a semi-functional ciphertext. In the next q games, individual responses to the key extraction queries are replaced one-by-one with semi-functional decryption keys. A change from normal to semi-functional (whether ciphertext or decryption keys) is not detectable to an adversary assuming that the decision linear problem is hard. The last change is to provide the adversary with a semi-functional ciphertext on a random element and the adversary does not notice this change if DBDH is assumed to be hard. This approach to security reduction has been called *dual system encryption* in [170] and has been used later to build several cryptographic primitives including a constant size ciphertext HIBE [131].

Applying this novel proof technique, Waters achieves two important goals. The first one is to construct an IBE whose security is based on a static assumption (i.e., the instance of the hard problem does not depend on some parameter of the scheme), does not use random oracles and for which the size of the public parameters is independent of the security parameter. Apart from the last point, the previous two points were already achieved in the previous IBE scheme of Waters [169]. Recall that in [169], the number of elliptic curve points in the public parameter is $n + 3$ to achieve $n/2$-bit security. The generalization in [60, 137] reduces this to a fraction of n, but, still the number of points in the public parameters grows with increase in n. The second achievement is to extend the IBE to an HIBE where the security degradation is not exponential in the number of levels.

A comment on the implementation aspects of Waters-2009 IBE is however, in order. For practical security levels such as 80-bit or 128-bit security, the actual size of the public parameters in [60] is comparable to that of Waters-2009. On the other hand, the efficiencies of encryption and decryption in [60] is much better than that of Waters-2009. So, even though, several new ideas are introduced in Waters-2009 IBE scheme, at the current point of time it is still mainly of theoretical interest. Below we describe the IBE scheme and the security proof.

The description of the scheme is quite involved and so is the security reduction. As a running intuition, it might help the reader to keep in mind that the purpose is to be able to change some parts of a normal ciphertext to obtain a semi-functional ciphertext and also change some parts of a normal decryption key to obtain a semi-functional decryption key. This essentially requires some kind of implicit splitting of keys and ciphertexts into two parts.

Set-Up: PKG chooses independent and random generators P, V, V_1, V_2, W, U, Q from G and independent and uniform random elements a_1, a_2, b, α from $\mathbb{Z}_p$. It sets $T_1 = V + a_1 V_1$ and $T_2 = V + a_2 V_2$. The public parameters consist of the following elements.

$$P, \ bP, \ a_1P, \ a_2P, \ ba_1P, \ ba_2P, \ T_1, \ T_2, \ bT_1, \ bT_2, \ W, \ U, \ Q, \ e(P,P)^{\alpha a_1 b}.$$

The master secret key consists of the following elements: $\alpha P, \ \alpha a_1 P, \ V, \ V_1, \ V_2$. Identities are elements of $\mathbb{Z}_p$.

If we consider $a = a_1 + a_2$, then a_1P and a_2P is a split of aP and ba_1P and ba_2P is a split of abP. It will be helpful to think of αP as the main component of the master secret key.[1]

Key-Gen The input is an identity $id \in \mathbb{Z}_p$. The PKG chooses independent and uniform random elements r_1, r_2, z_1, z_2, ktag from $\mathbb{Z}_p$ and sets $r = r_1 + r_2$. The secret key d_{id} for the identity id is defined to be $(D_1, \ldots, D_7, K, ktag)$, where

$$D_1 = \alpha a_1 P + rV; \quad D_2 = -\alpha P + rV_1 + z_1 P; \quad D_3 = -z_1(bP);$$
$$D_4 = rV_2 + z_2 P; \quad D_5 = -z_2(bP); \quad D_6 = r_2 bP;$$
$$D_7 = r_1 P; \quad K = r_1(id\, U + ktag\, W + Q).$$

The elements D_1, D_2, D_4 depend on $r = r_1 + r_2$, whereas D_6 depends on r_2 and D_7 and K depend on r_1. Note that, K is the only element of the key which depends on the identity id.

Encrypt: The input is an identity $id \in \mathbb{Z}_p$ and a message $M \in G_T$. The sender chooses independent and uniform random s_1, s_2, t, ctag from $\mathbb{Z}_p$ and sets $s = s_1 + s_2$. The ciphertext is $(C_0, \ldots, C_7, E_1, E_2, ctag)$ where

$$C_0 = M \times (e(P,P)^{\alpha a_1 b})^{s_2}; \quad C_1 = s(bP); \quad C_2 = s_1(ba_1 P);$$
$$C_3 = s_1(a_1 P); \quad C_4 = s_2(ba_2 P); \quad C_5 = s_2(a_2 P);$$
$$C_6 = s_1 T_1 + s_2 T_2; \quad C_7 = s_1(bT_1) + s_2(bT_2) - tW;$$
$$E_1 = t(id\, U + ctag\, W + Q); \quad E_2 = tP.$$

The message M is masked by $(e(P,P)^{\alpha a_1 b})^{s_2}$, where s_2 is the randomiser and $e(P,P)^{\alpha a_1 b}$ is part of the public parameters. C_0 is the only component of the ciphertext which depends on α. The value s is formed as $s = s_1 + s_2$. C_1 depends on s; C_2, C_3 depend on s_1; C_4, C_5 depend on s_2 and C_6, C_7 depend on both s_1 and s_2. The only part of the ciphertext which depends on the identity is E_1. Element $E_2 = tP$ and the other elements affected by t are C_7 and E_1.

Decrypt: The input is a ciphertext $(C_0, \ldots, C_7, E_1, E_2, ctag)$, an identity id and a decryption key $d_{id} = (D_1, \ldots, D_7, K, ktag)$. If $ctag = ktag$, then the ciphertext cannot be decrypted and this event occurs with probability $1/p$. The decryption consists of several computations. (Note $r = r_1 + r_2$ and $s = s_1 + s_2$.)

$$A_1 = e(C_1, D_1) \times e(C_2, D_2) \times e(C_3, D_3) \times e(C_4, D_4) \times e(C_5, D_5)$$
$$= e(P,P)^{\alpha a_1 b(s_1 + s_2)} \times e(P,V)^{(s_1 + s_2)rb} \times e(P,P)^{-\alpha a_1 bs_1} \times e(P,V_1)^{a_1 bs_1 r}$$
$$\times e(P,P)^{ba_1 s_1 z_1} \times e(P,P)^{-ba_1 s_1 z_1} \times e(P,V_2)^{ba_2 s_2 r}$$
$$\times e(P,P)^{ba_2 s_2 z_2} \times e(P,P)^{-ba_2 s_2 z_2}$$
$$= e(P,P)^{\alpha a_1 bs_2} \times e(V,P)^{b(s_1 + s_2)r} \times e(V_1,P)^{a_1 bs_1 r} \times e(V_2,P)^{a_2 bs_2 r},$$

---

[1] In both the conference version [170] and the full version [171], the element P is shown to be part of the master secret key. If this is the case, then the element E_2 of the ciphertext (see Encrypt) cannot be generated. We believe this is an error and P should be part of the public parameters.

$$A_2 = e(C_6, D_6) \times e(C_7, D_7)$$
$$= e(V,P)^{b(s_1+s_2)r} \times e(V_1,P)^{a_1 bs_1 r} \times e(V_2,P)^{a_2 bs_2 r} \times e(P,W)^{-r_1 t},$$

$$A_3 = A_1/A_2 = e(P,P)^{\alpha a_1 bs_2} \times e(P,W)^{r_1 t}.$$

If ctag $\neq$ ktag, then

$$A_4 = \left(\frac{e(E_1, D_7)}{e(E_2, K)} \right)^{1/(\text{ctag}-\text{ktag})} = e(P,W)^{r_1 t},$$

$$\frac{A_3}{A_4} = e(P,P)^{\alpha a_1 bs_2}.$$

Finally, the message is obtained as $M = C_0/(A_3/A_4)$.

The above completes the description of the scheme. We now turn to the security analysis. For this, it is first necessary to introduce the notions of semi-functional ciphertexts and decryption keys.

Semi-functional ciphertexts.

Let $(C_0, \ldots, C_7, E_1, E_2, \text{ctag})$ be a ciphertext generated by the encryption algorithm. Choose a uniform random x in $\mathbb{Z}_p$ and define a semi-functional ciphertext $(C_0', \ldots, C_7', E_1', E_2', \text{ctag})$ as follows.

$C_i' = C_i$ for $i = 0, 1, 2, 3$; $E_1' = E_1$ and $E_2' = E_2$;
$C_4' = C_4 + ba_2 xP$; $C_5' = C_5 + a_2 xP$; $C_6' = C_6 + a_2 xV_2$; $C_7' = C_7 + a_2 bxV_2$.

The components C_0, C_1, C_2, C_3 and E_1, E_2 are left unchanged. The four components C_4 to C_7 are changed. The change corresponds to a separate (partial) randomisation by the value x. In the above, the role played by s_2 in the normal ciphertext is now played by $(s_2 + x)$ while the roles of s_1 and s remain unchanged. In other words, the randomising triplet (s, s_1, s_2) is changed to $(s, s_1, s_2 + x)$. This is easy to see for C_4 and C_5 and can be seen for C_6 and C_7 by expanding T_1 and T_2 in terms of V_1 and V_2 as defined during the set-up. Note that changing (s, s_1, s_2) to $(s' = s, s_1' = s_1, s_2' = s_2 + x)$ violates the condition $s' = s_1' + s_2'$ which would have to be true for a proper ciphertext.

The modification to C_4 and C_5 can be done using $ba_2 P$ and $a_2 P$ which form part of the public parameters and hence can be done by anybody. However, the changes to C_6 and C_7 require the use of the secret element V_2 apart from the values a_2 and b. V_2 is part of the master secret key, whereas a_2 and b are not. So, even the master secret key is not sufficient to form a semi-functional ciphertext. However, these are not of any concern in the actual scheme. On the other hand, in the security reduction we will see that the simulator can generate a semi-functional ciphertext because it will have the values a_2 and b.

Semi-functional decryption keys.

Let $(D_1, \ldots, D_7, K, \text{ktag})$ be a decryption key for an identity id. Choose a uniform random γ from $\mathbb{Z}_p$ and define a semi-functional decryption key

$$(D_1', \ldots, D_7', K, \text{ktag}) \tag{8.2.1}$$

as follows.

$D_i' = D_i$ for $i = 3$ and $i = 5, 6, 7$;
$D_1' = D_1 - a_1 a_2 \gamma P; \ D_2' = D_2 + a_2 \gamma P; \ D_4' = D_4 + a_1 \gamma P.$

Only the elements D_1, D_2 and D_4 are modified and all other elements of the normal decryption key remain unchanged. The changed elements depend on the randomiser γ. Modifications to D_2 and D_4 can be done based on public knowledge ($a_2 P$ and $a_1 P$) and γ. But, the modification to D_1 requires $a_1 a_2 P$ which cannot be computed from the public information P, $a_1 P$ and $a_2 P$ (unless CDH is easy). In fact, it is also not possible to compute this from the master secret key. However, the simulator in the proof will have a_1 and a_2 and will be able to perform this computation.

A decryption of a semi-functional ciphertext with a normal key will succeed because of the following computation.

$$\frac{e(ba_2 xP, D_4) \times e(a_2 xP, D_5)}{e(a_2 xV_2, D_6) \times e(a_2 bxV_2, D_7)} = 1$$

when D_4, D_5, D_6 and D_7 come from a normally generated private key. Similarly, the decryption of a normal ciphertext with a semi-functional key will succeed due to the following computation.

$$e(C_1, -a_1 a_2 \gamma P) \times e(C_2, a_2 \gamma P) \times e(C_4, a_1 \gamma P) = 1.$$

If on the other hand, an attempt is made to decrypt a semi-functional ciphertext with a semi-functional decryption key, then the recovered value will be the message times the quantity $e(P, P)^{-a_1 a_2 x \gamma b}$ and will be a random value due to the randomness of x (and γ).

Theorem 8.2. *The IBE scheme described above is $(t, \varepsilon_{ibe}, q)$-CPA-secure assuming that DLIN is (t', ε_{dlin})-hard and DBDH is (t', ε_{dbdh})-hard, where*

$$\varepsilon_{ibe} \leq (q+1)\varepsilon_{dlin} + \varepsilon_{dbdh}$$

and $t' = t + O(q\tau)$ and τ is the time required for a scalar multiplication in G.

Proof : The proof proceeds through a total of $(q+3)$ games. The initial game Game_{real} is the actual security game used in defining CPA-security of IBE. Then there are $(q+1)$ security games Game_0 to Game_q followed by the Game_{final}. Suppose $X_{real}, X_0, \ldots, X_q, X_{final}$ be the events that the adversary's guesses in games

Game_{real}, Game_0 to Game_q and Game_{final} respectively are correct. The changes between the games are as follows.

1. The change between Game_{real} and Game_0 is that the challenge ciphertext is changed from normal to semi-functional.
2. The change between Game_{k-1} and Game_k is that the reply to the k-th key-extraction query is changed from normal to semi-functional. The replies to queries numbered 1 to $k-1$ are semi-functional keys and the replies to queries numbered $k+1$ to q are normal keys.
3. The change between Game_q and Game_{final} is that the challenge ciphertext is a semi-functional ciphertext on a random element of G_T. This game statistically hides the simulator's uniform random choice of one of the two messages and so $\Pr[X_{final}] = 1/2$.

A sequence of lemmas below shows the following results.

1. $\Pr[X_{real}] - \Pr[X_0] \leq \varepsilon_{dlin}$.
2. $\Pr[X_{k-1}] - \Pr[X_k] \leq \varepsilon_{dlin}$ for $k = 1, \ldots, q$.
3. $\Pr[X_q] - \Pr[X_{final}] \leq \varepsilon_{dbdh}$.

The probability of X_{final} is $1/2$ and so

$$\varepsilon_{ibe} = \left| \Pr[X_{real}] - \frac{1}{2} \right| = |\Pr[X_{real}] - \Pr[X_{final}]|$$

$$= |\Pr[X_{real}] - \Pr[X_0]| + \sum_{k=1}^{q} |\Pr[X_{k-1}] - \Pr[X_k]|$$

$$+ |\Pr[X_q] - \Pr[X_{final}]|$$

$$\leq (q+1)\varepsilon_{dlin} + \varepsilon_{dbdh}.$$

So, the task is to prove the above statements on the indistinguishability of two successive games (under a complexity assumption). This is done in the three lemmas below. In the proofs of these lemmas, $\mathscr{A}$ is a CPA-adversary against the Waters-2009 IBE and $\mathscr{B}$ is an algorithm which interacts with $\mathscr{A}$ to solve either DLIN or DBDH problem.

Lemma 8.1. $|\Pr[X_{real}] - \Pr[X_0]| \leq \varepsilon_{dlin}$.

Proof : The input to $\mathscr{B}$ is an instance $(P, R, S, c_1 P, c_2 R, T)$ of DLIN where T is either $(c_1 + c_2)S$ or a random element of G. $\mathscr{B}$ sets up the IBE scheme, answers decryption queries and generates challenge ciphertext. In Game_{real}, this is a normal ciphertext while in Game_0, this is a semi-functional ciphertext. The adversary's ability to distinguish between these two types of ciphertexts translates into the ability of $\mathscr{B}$ to determine whether T is real or random. Now we describe how $\mathscr{B}$ simulates the protocol environment for $\mathscr{A}$.

Set-Up: $\mathscr{B}$ chooses independent and uniform random elements b, α, y, y_1, y_2 from $\mathbb{Z}_p$ and independent and uniform random elements U, W, Q from G. It then sets $a_1 P$ to be equal to R and $a_2 P$ to be equal to S. Note that $\mathscr{B}$ cannot actually determine a_1 or a_2 (without solving discrete log in G). Then, $\mathscr{B}$ computes

$$bP, b(a_1 P) = bR, b(a_2 P) = bS, V = yP, V_1 = y_1 P, V_2 = y_2 P,$$
$$T_1 = V + a_1 V_1 = V + y_1(a_1 P) = V + y_1 R,$$
$$T_2 = V + a_2 V_2 = V + y_2(a_2 P) = V + y_2 S,$$
$$bT_1, bT_2,$$
$$e(P,P)^{\alpha a_1 b} = e(P, a_1 P)^{\alpha b} = e(P,R)^{\alpha b}.$$

So, $\mathscr{B}$ can provide $\mathscr{A}$ with a proper set of public parameters. The master secret key is $(\alpha P, \alpha(a_1 P), V, V_1, V_2) = (\alpha P, \alpha R, V, V_1, V_2)$ which $\mathscr{B}$ can compute.

Key Extraction Query: Since $\mathscr{B}$ has the master key, it can generate normal secret keys for any identity of the adversary's choice.

Challenge: $\mathscr{B}$ receives two messages M_0 and M_1 and a challenge identity id* and chooses a uniform random bit β. First $\mathscr{B}$ generates a normal ciphertext

$$(C_0, \ldots, C_7, E_1, E_2, \text{ctag})$$

for M using the encryption algorithm. Let s_1, s_2 and t be the random elements of $\mathbb{Z}_p$ used in creating this ciphertext. This ciphertext is converted into a ciphertext $(C_0', \ldots, C_7', E_1, E_2, \text{ctag})$ as follows:

$$C_0' = C_0 \times (e(c_1 P, R) \times e(P, c_2 R))^{b\alpha}, \ C_1' = C_1 + b(c_1 P), \ C_2' = C_2 - b(c_2 R),$$
$$C_3' = C_3 - c_2 R, \ C_4' = C_4 + bT, \ C_5' = C_5 + T,$$
$$C_6' = C_6 + y(c_1 P) - y_1(c_2 R) + y_2 T, \ C_7' = C_7 + b(y(c_1 P) - y_1(c_2 R) + y_2 T).$$

$(C_0', \ldots, C_7', E_1, E_2, \text{ctag})$ is returned to $\mathscr{A}$ as the challenge ciphertext.

If $T = (c_1 + c_2)S$, then the challenge is a normal ciphertext under the implicit assignment $s_1' = s_1 - c_2$ and $s_2' = s_2 + c_1 + c_2$. If, on the other hand, T is a uniform random element of G, then the challenge is a semi-functional ciphertext. For ease of understanding we work out the details. Let $s_1' = s_1 - c_2$, $s_2' = s_2 + c_1 + c_2$ and $s' = s_1 + s_2 + c_1$. Also, in the computation, we assume that T is real, i.e., $T = (c_1 + c_2)S$. The case of random T is similar and is mentioned later.

$$C_0' = C_0 \times (e(c_1 P, R) \times e(P, c_2 R))^{b\alpha}$$
$$= M_\beta \times e(P,R)^{\alpha b s_2} \times e(P,R)^{(c_1 + c_2)\alpha b}$$
$$= M_\beta \times e(P,R)^{\alpha b(s_2 + c_1 + c_2)}$$
$$= M_\beta \times e(P,R)^{\alpha b s_2'},$$
$$C_1' = C_1 + b(c_1 P) = (s_1 + s_2)bP + bc_1 P = s'(bP),$$
$$C_2' = C_2 - b(c_2 R) = s_1(bR) - bc_2 R = s_1'(bR),$$
$$C_3' = C_3 - c_2 R = s_1 R - c_2 R = s_1' R,$$
$$C_4' = C_4 + bT = s_2(bS) + bT = b(s_2 S + T) = s_2'(bS),$$

$$C_5' = C_5 + T = s_2 S + T = s_2' S,$$
$$C_6' = C_6 + y(c_1 P) - y_1(c_2 R) + y_2 T$$
$$= s_1 T_1 + s_2 T_2 + y(c_1 P) - y_1(c_2 R) + y_2 T$$
$$= s_1(V + y_1 R) + s_2(V + y_2 S) + y(c_1 P) - y_1(c_2 R) + y_2 T$$
$$= s_1(V + y_1 R) + s_2(V + y_2 S) - y c_2 P - y_1 c_2 R + y c_1 P + y c_2 P + y_2(c_1 + c_2) S$$
$$= (s_1 - c_2) T_1 + (s_2 + c_1 + c_2) T_2$$
$$= s_1' T_1 + s_2' T_2,$$
$$C_7' = C_7 + b(y(c_1 P) - y_1(c_2 R) + y_2 T)$$
$$= s_1(b T_1) + s_2(b T_2) - t W + b(y(c_1 P) - y_1(c_2 R) + y_2 T)$$
$$\cdot\ \cdots$$
$$= s_1'(b T_1) + s_2'(b T_2) - t W.$$

When T is random, we can write $T = (c_1 + c_2) S + a_2 x P$ for some random x. The reader can now verify that this gives a semi-functional ciphertext.

Guess: At the end of the game, $\mathscr{A}$ returns its guess β' and $\mathscr{B}$ returns $1 \oplus \beta \oplus \beta'$.

In the above simulation, if T is real then $\mathscr{B}$ is simulating Game_{Real} and if T is random then $\mathscr{B}$ is simulating Game_0. A crucial point to note is that, $\mathscr{B}$ can generate a normal private key for the challenge identity but not a semi-functional private key. Hence, $\mathscr{B}$ cannot by itself decide whether the challenge ciphertext is semifunctional or normal. If, on the other hand, $\mathscr{A}$ can distinguish between the two, then the advantage of $\mathscr{A}$ translates into the advantage of $\mathscr{B}$ to solve the given instance of DLIN problem. □

Lemma 8.2. *For* $1 \le k \le q$, $|\Pr[X_{k-1}] - \Pr[X_k]| \le \varepsilon_{dlin}$.

Proof : As in Lemma 8.1, the input to $\mathscr{B}$ is an instance $(P, R, S, c_1 P, c_2 R, T)$ of the DLIN problem. Based on this $\mathscr{B}$ simulates the protocol environment as follows.

Set-Up: $\mathscr{B}$ chooses independent and uniform random values $\alpha, a_1, a_2, y_1, y_2, w, u, h$ and computes the following quantities.

$$bP = R, \ b a_1 P = a_1 R, \ b a_2 P = a_2 R, \ V = -a_1 a_2 S,$$
$$V_1 = a_2 S + y_1 P, \ V_2 = a_1 S + y_2 P, \ e(P, P)^{\alpha a_1 b} = e(R, P)^{\alpha a_1},$$
$$T_1 = V + a_1 V_1 = -a_1 a_2 S + a_1 a_2 S + a_1 y_1 P = a_1 y_1 P,$$
$$T_2 = V + a_2 V_1 = a_2 y_2 P,$$
$$b T_1 = V + a_1 V_1 = y_1 a_1 R, \ b T_2 = V + a_2 V_1 = a_2 y_2 R.$$

The computation of the expressions for T_2, $b T_1$ and $b T_2$ are similar to that of T_1.

$\mathscr{B}$ chooses independent and uniform random γ_1, γ_2 in $\mathbb{Z}_p$ and sets $W = R + w P$, $U = -\gamma_1 R + u P$ and $Q = -\gamma_2 R + h P$. $\mathscr{B}$ sets the public parameters as per the protocol and gives them to $\mathscr{A}$. Since α is known to $\mathscr{B}$, it also knows the master secret key.

For any identity id, define $F(\mathsf{id}) = \gamma_1 \mathsf{id} + \gamma_2$. This F is a pairwise independent function and so if the adversary is given $F(\mathsf{id})$ for some identity id, then for some $\mathsf{id}' \ne \mathsf{id}$, $F(\mathsf{id}')$ is uniformly distributed over $\mathbb{Z}_p$.

Key Extraction Query: There are a total of q queries covering both the phases. The simulation depends on the query number and does not depend on whether it is a Phase 1 or a Phase 2 query. Suppose the i-th query is to be simulated. Depending on the value of i, there are three cases.

Case $i > k$. $\mathscr{B}$ generates a normal key for id using the master secret key that it knows. This key is returned to $\mathscr{A}$.

Case $i < k$. $\mathscr{B}$ creates a normal decryption key and then converts it into a semi-functional key using the method of generating semi-functional keys (see equations mentioned after (8.2.1)). Since $\mathscr{B}$ knows a_1 and a_2 this can be done. The semi-functional key is returned.

Case $i = k$. $\mathscr{B}$ sets $\mathsf{ktag}^* = F(\mathsf{id})$ and using this ktag, it runs the key generation algorithm to obtain a normal decryption key $d_{\mathsf{id}} = (D_1, \ldots, D_7, K, \mathsf{ktag}^*)$ for id. Suppose the random values used to generate this key are r_1, r_2, z_1 and z_2. $\mathscr{B}$ then defines

$$D'_1 = D_1 - a_1 a_2 T, \ D'_2 = D_2 + a_2 T + y_1(c_1 P),$$
$$D'_3 = D_3 + y_1(c_2 R), \ D'_4 = D_4 + a_1 T + y_2(c_1 P),$$
$$D'_5 = D_5 + y_2(c_2 R), \ D'_6 = D_6 + c_2 R,$$
$$D'_7 = D_7 + (c_1 P), \ K' = K + (u \ \mathsf{id} + h + w \ \mathsf{ktag}^*)(c_1 P).$$

The crucial point is that $\mathsf{ktag}^* = F(\mathsf{id}) = \gamma_1 \mathsf{id} + \gamma_2$ allows K' to be properly created. Since,

$$\begin{aligned}
K' &= r_1(\mathsf{id}U + \mathsf{ktag}^* W + Q) + (u \ \mathsf{id} + h + w \ \mathsf{ktag}^*)(c_1 P) \\
&= r_1(-\gamma_1 \mathsf{id}R + u\mathsf{id}P + \mathsf{ktag}^*(R + wP) - \gamma_2 R + hP) + c_1(u \ \mathsf{id} + h + w \ \mathsf{ktag}^*)P \\
&= (r_1 + c_1)(u\mathsf{id}P + w\mathsf{ktag}^* P + hP) \\
&= r'_1(\mathsf{id}U + \mathsf{ktag}^* W + Q).
\end{aligned}$$

Also, the definition of D'_2 and D'_4 implicitly sets $z'_1 = z_1 - y_1 c_1$ and $z'_2 = z_2 - y_2 c_2$.

If $T = (c_1 + c_2)S$ and we set $r'_1 = r_1 + c_1$ and $r'_2 = r_2 + c_2$, then the key $(D'_1, \ldots, D'_7, K, \mathsf{ktag}^*)$ is a normal key and if T is a uniform random element of G, then it is a semi-functional key under the use of the randomness r'_1 and r'_2. We provide the details of this computation for D'_1, D'_2 and D'_4 for the case T is real, i.e., $T = (c_1 + c_2)S$. Also, let $r'_1 = r_1 + c_1, r'_2 = r_2 + c_2, z'_1 = z_1 - y_1 c_1$ and $z'_2 = z_2 - y_2 c_2$.

$$\begin{aligned}
D'_1 &= D_1 - a_1 a_2 T \\
&= \alpha a_1 P + (r_1 + r_2)V - a_1 a_2 T \\
&= \alpha a_1 P - a_1 a_2(r_1 + r_2 + c_1 + c_2)S \\
&= \alpha a_1 P + (r'_1 + r'_2)(-a_1 a_2 S) \\
&= \alpha a_1 P + (r'_1 + r'_2)V, \\
D'_2 &= D_2 + a_2 T + y_1(c_1 P) \\
&= -\alpha P + (r_1 + r_2)V_1 + z_1 P + a_2(c_1 + c_2)S + y_1(c_1 P) \\
&= -\alpha P + (r_1 + r_2)(a_2 S + y_1 P) + c_1(a_2 S + y_1 P) + c_2(a_2 S + y_1 P) - c_2 y_1 P + z_1 P \\
&= -\alpha P + (r_1 + c_1 + r_2 + c_2)V_1 + (z_1 - c_1 y_1)P \\
&= -\alpha P + (r'_1 + r'_2)V_1 + z'_1 P,
\end{aligned}$$

$$D'_4 = (r_1 + r_2)V_2 + z_2P + a_1(c_1 + c_2)S + y_2(c_1P)$$
$$= (r_1 + r_2)V_2 + c_1(a_1S + y_2P) + c_2(a_1S + y_2P) + z_2P - c_2y_2P$$
$$= (r'_1 + r'_2)V_2 + z'_2P.$$

If T is random, we can write $T = (c_1 + c_2)S + \gamma P$ for some (unknown) $\gamma \in \mathbb{Z}_p$, which gives a semi-functional key.

Challenge: The challenge identity is id^* and the two messages are M_0 and M_1. $\mathscr{B}$ has to return a semi-functional ciphertext. (From Game$_0$ onwards the challenge ciphertext is to be a semi-functional ciphertext.) Since $\mathscr{B}$ does not have the element bV_2 (recall that $\mathscr{B}$ sets $bP = R$ in the public parameter and does not know b), it cannot directly create a semi-functional ciphertext. The method of doing this is described below.

$\mathscr{B}$ sets $\text{ctag}^* = F(\text{id}^*)$. As is usual $\mathscr{B}$ generates a uniform random bit β and sets $M^* = M_\beta$. $\mathscr{B}$ now obtains a normal ciphertext $(C_0, C_1, \ldots, C_7, E_1, E_2, \text{ctag}^*)$ for M^* under identity id^* and let s_1, s_2 and t be the random values used to generate it. $\mathscr{B}$ now chooses a uniform random x in $\mathbb{Z}_p$ and computes the following.

$$C'_4 = C_4 + xa_2R, \; C'_5 = C_5 + xa_2P,$$
$$C'_6 = C_6 + xa_2V_2, \; C'_7 = C_7 + y_2xa_2R - a_1xwa_2S,$$
$$E'_1 = E_1 + a_1a_2x(\text{id}^* u + h + \text{ctag}^* w)S, \; E'_2 = E_2 + a_1a_2xS.$$

The semi-functional ciphertext is

$$(C_0, C_1, C_2, C_3, C'_4, \ldots, C'_7, E'_1, E'_2, \text{ctag}^*).$$

Elements C'_4, C'_5 and C'_6 are generated by the usual method of generating a semi-functional ciphertext. Justification for C'_7, E'_1 and E'_2 are given by the following computation. Let $t' = t + \rho a_1 a_2 x$ and $S = \rho P$, then

$$C'_7 = C_7 + y_2xa_2(bP) - a_1xwa_2S$$
$$= C_7 + y_2xa_2bP - a_1xwa_2S + a_1a_2bxS - a_1a_2bxS$$
$$= C_7 + a_2bx(y_2P + a_1S) - a_1a_2x(b + w)S$$
$$= (bs_1T_1 + bs_2T_2 - tW) + a_2bxV_2 - a_1a_2x(b + w)S$$
$$= (bs_1T_1 + bs_2T_2 - t'W) + a_2bxV_2.$$

The last equality follows from the following:

$$-tW - a_1a_2x(b + w)S = -t(R + wP) - a_1a_2x(b + w)S$$
$$= -tbP - twP - a_1a_2x(b + w)(\rho P)$$
$$= (-t - \rho a_1a_2x)(bP) + (-t - \rho a_1a_2x)(wP)$$
$$= (-t - \rho a_1a_2x)(R + wP)$$
$$= (-t - \rho a_1a_2x)W$$
$$= -t'W.$$

In other words, $\mathscr{B}$ implicitly sets $t'P = tP + a_1 a_2 xS$ for some unknown t'. This could be a problem in the computation of E_1'. Setting $\mathrm{ctag}^* = F(\mathrm{id}^*)$ the simulator can avoid this problem as shown below.

$$
\begin{aligned}
E_1' &= E_1 + a_1 a_2 x(\mathrm{id}^*\, u + h + \mathrm{ctag}^*\, w)S \\
&= t(\mathrm{id}^*\, U + \mathrm{ctag}^*\, W + Q) + a_1 a_2 x(\mathrm{id}^*\, u + h + \mathrm{ctag}^*\, w)S \\
&= t(\mathrm{id}^*(-\gamma_1 R + uP) + (\gamma_1\, \mathrm{id}^* + \gamma_2)(R + wP) - \gamma_2 R + hP) \\
&\quad + a_1 a_2 x(\mathrm{id}^*\, u + h + \mathrm{ctag}^*\, w)S \\
&= t(\mathrm{id}^*\, u + h + \mathrm{ctag}^*\, w)P + a_1 a_2 x(\mathrm{id}^*\, u + h + \mathrm{ctag}^*\, w)\rho P \\
&= (\mathrm{id}^*\, u + h + \mathrm{ctag}^*\, w)(tP + a_1 a_2 x\rho P) \\
&= t'(\mathrm{id}^*\, u + h + \mathrm{ctag}^*\, w)P \\
&= t'(\mathrm{id}^*\, U + \mathrm{ctag}^*\, W + Q), \\
E_2' &= E_2 + a_1 a_2 xS \\
&= tP + a_1 a_2 x\rho P \\
&= t'P.
\end{aligned}
$$

However, $\mathscr{B}$ cannot use the above strategy to generate a semifunctional ciphertext for the kth identity and check whether the corresponding key is semi-functional or not. In that case the decryption will fail unconditionally as ctag will be equal to ktag.

Guess: $\mathscr{A}$ outputs its guess β' and $\mathscr{B}$ outputs $1 \oplus \beta \oplus \beta'$.

If T is real (i.e., $T = (c_1 + c_2)S$), then the output of the k-th query is a normal key while if T is random, then the output of the k-th query is a semi-functional key. In other words, if T is real, then $\mathscr{A}$ is playing Game_{k-1}, and if T is random, then $\mathscr{A}$ is playing Game_k. This proves the lemma. $\square$

Lemma 8.3. $|\Pr[X_q] - \Pr[X_{final}]| \leq \varepsilon_{dbdh}$.

Proof : In both Game_q and Game_{final}, the challenge ciphertexts and all the decryption keys are semi-functional. However, in Game_{final} the adversary gets a ciphertext corresponding to a random element of G_T. The input to $\mathscr{B}$ is a DBDH instance $(P, c_1 P, c_2 P, c_3 P, T)$ and the task is to determine whether $T = e(P,P)^{c_1 c_2 c_3}$ or whether T is a uniform random element of G_T. $\mathscr{B}$ simulates the protocol environment as follows.

Set-Up: $\mathscr{B}$ chooses independent and uniform random values $a_1, b, y, y_1, y_2, w, u, h$ from $\mathbb{Z}_p$ and computes the following.

$bP,\ a_1 P,\ a_2 P = c_2 P,\ ba_1 P,\ ba_2 P = b(c_2 P),\ V = yP,\ V_1 = y_1 P,\ V_2 = y_2 P,$
$W = wP,\ U = uP,\ Q = hP,\ e(P,P)^{a_1 \alpha b} = e(c_1 P, c_2 P)^{a_1 b},$
$T_1 = V + a_1 V_1,\ T_2 = V + y_2(c_2 P),\ bT_1\ \text{and}\ bT_2.$

The public parameters are given to $\mathscr{A}$. Note that $\mathscr{B}$ implicitly sets $\alpha = c_1 c_2$ and $a_2 = c_2$. So the components αP and $\alpha a_1 P$ of the master secret key are not available with $\mathscr{B}$.

Key Extraction Query: The generation of a semi-functional decryption key for an identity id is done as follows. $\mathcal{B}$ chooses independent and uniform random values $r_1, r_2, z_1, z_2, \gamma'$, ktag from $\mathbb{Z}_p$. Let $r = r_1 + r_2$ and the idea is to implicitly set the variable γ to be equal to $c_1 + \gamma'$. The semi-functional key is created as follows.

$$D_1 = -(\gamma' a_1)(c_2 P) + rV, \; D_2 = \gamma'(c_2 P) + rV_1 + z_1 P, \; D_3 = -z_1(bP),$$
$$D_4 = a_1(c_1 P) + (a_1 \gamma')P + rV_2 + z_2 P,$$
$$D_5 = -z_2(bP), \; D_6 = r_2(bP), \; D_7 = r_1 P, \; K = r_1(\text{id } U + \text{ktag } W + Q).$$

We show that the derivation of D_1, D_2 and D_4 gives a semi-functional key. All other components of the private key are generated according to the original key generation algorithm. Note $\alpha = c_1 c_2$, $c_2 = a_2$ and set $\gamma = c_1 + \gamma'$.

$$\begin{aligned}
D_1 &= -\gamma' a_1(c_2 P) + rV \\
&= \alpha a_1 P + rV - \alpha a_1 P - \gamma' a_1(c_2 P) \\
&= (\alpha a_1 P + rV) - a_1 c_2(c_1 + \gamma')P \\
&= (\alpha a_1 P + rV) - a_1 a_2 \gamma P, \\
D_2 &= \gamma'(c_2 P) + rV_1 + z_1 P \\
&= (-\alpha P + rV_1 + z_1 P) + \alpha P + \gamma' c_2 P \\
&= (-\alpha P + rV_1 + z_1 P) + c_2(c_1 + \gamma')P \\
&= (-\alpha P + rV_1 + z_1 P) + a_2 \gamma P, \\
D_4 &= a_1(c_1 P) + (a_1 \gamma')P + rV_2 + z_2 P \\
&= (rV_2 + z_2 P) + a_1(c_1 + \gamma')P \\
&= (rV_2 + z_2 P) + a_1 \gamma P.
\end{aligned}$$

Challenge: $\mathcal{A}$ provides id* and two messages M_0 and M_1. $\mathcal{B}$ generates a ciphertext which is semi-functional and either encrypts M_β or encrypts a uniform random element of G_T according as T is real or random. Here β is a uniform random bit chosen by $\mathcal{B}$.

$\mathcal{B}$ starts by choosing independent and uniform random s_1, t, x' and ctag from $\mathbb{Z}_p$. The idea is to implicitly set s_2 to be equal to c_3 and x to be equal to $-c_3 + x'$. Next, the following computations are performed to generate the challenge ciphertext.

$$\begin{aligned}
&C_0 = M_\beta T^{a_1 b}, \\
&C_1 = s_1 bP + b(c_3 P), \; C_2 = ba_1 s_1 P, \; C_3 = a_1 s_1 P, \; C_4 = x'b(c_2 P), \; C_5 = x'(c_2 P), \\
&C_6 = s_1 T_1 + y(c_3 P) + (y_2 x')(c_2 P), \; C_7 = s_1(bT_1) + yb(c_3 P) + (y_2 x' b)(c_2 P) - tW, \\
&E_1 = t(\text{id } U + \text{ctag } W + Q), \; E_2 = tP.
\end{aligned}$$

The components C_1, C_2, C_3 and E_1, E_2 are generated by the usual encryption process. The computation showing the other elements is given below. Assume $a_2 = c_2$, $x' = x + c_3$ and $s_2 = c_3$.

$$\begin{aligned}
C_4 &= bx'(c_2 P) = ba_2(s_2 + x)P, \\
C_5 &= x'(c_2 P) = a_2(s_2 + x)P, \\
C_6 &= s_1 T_1 + y(c_3 P) + (y_2 x')(c_2 P)
\end{aligned}$$

$$= s_1 T_1 + y c_3 P + y_2 (x + c_3) c_2 P$$
$$= s_1 T_1 + y c_3 P + y_2 c_2 x P + c_2 c_3 y_2 P$$
$$= s_1 T_1 + c_3 (y P + y_2 c_2 P) + c_2 x (y_2 P)$$
$$= s_1 T_1 + s_2 (V + y_2 c_2 P) + a_2 x V_2$$
$$= s_1 T_1 + s_2 T_2 + a_2 x V_2,$$
$$C_7 = s_1 (b T_1) + y b (c_3 P) + y_2 x' b (c_2 P) - t W$$
$$= b (s_1 T_1 + y c_3 P + y_2 x' c_2 P) - t W.$$

In the above calculation C_7 is derived in the same way as C_6. Finally, if $T = e(P,P)^{c_1 c_2 c_3}$, then

$$C_0 = M_\beta \times \left(e(P,P)^{\alpha a_1 b} \right)^{s_2}.$$

Guess. At some point $\mathscr{A}$ outputs its guess β' and $\mathscr{B}$ returns $1 \oplus \beta \oplus \beta'$.

Note that, $\mathscr{B}$ cannot generate a normal private key for id^* and hence has to depend upon the adversary's guess. This way the advantage of $\mathscr{A}$ to distinguish between Game_q and Game_{final} translates into the advantage of $\mathscr{B}$ to solve the DBDH problem. $\square$

8.2.1 Extensions

The technique of dual system encryption used in the IBE scheme above has been used for the construction of several other primitives. We briefly mention some of these below.

1. A HIBE scheme has been constructed in [170]. The main novelty of this scheme is that it avoids the exponential security degradation in the depth of the HIBE which is true for the HIBE schemes we have seen so far in this book. This HIBE construction is not the first one to achieve this property. It was earlier achieved in [92]. But, the construction in [92] is very complicated and it is based on a non-static and non-standard security assumption. In comparison, the HIBE scheme in [170] is simpler and is based on simple static assumptions.
2. The paper [170] describes a broadcast encryption scheme which satisfies adaptive security, i.e., the target set of recipients to be attacked by the adversary can be chosen by the adversary during the course of the attack and need not be specified at the outset. For this application also, the construction in [170] provides the best known solution.
3. Later work by Lewko and Waters [131] have shown how to use dual-system encryption to obtain constant-size ciphertext HIBE. But, for this they have to work in composite order groups.

8.3 Conclusion

This chapter describes two important techniques to construct IBE in the adaptive-ID model without random oracle. Both approaches are distinct from the partitioning approach used to construct IBE schemes described in the previous chapters. The first one, due to Gentry, achieves a tight security reduction at the expense of a non-standard and non-static complexity assumption. The second one, attributed to Waters, introduces a novel paradigm to argue the security of cryptographic schemes. Applied in the context of IBE, this technique gives a construction where the public parameters, private keys and ciphertexts are constant number of group elements and security is based on DBDH and DLIN assumptions. Extending to HIBE, this technique provides a construction where the security does not degrade exponentially with the number of levels.

Chapter 9
IBE Without Pairing

Bilinear pairings on elliptic curve groups have been extensively used for the construction of identity-based encryption schemes and related primitives. This, however, does not give the complete picture about IBE schemes. When Shamir [155] had proposed the idea in 1984, he had been unable to give a solution to the problem. In the public domain, the problem remained open for about 17 years till the publication of the Boneh-Franklin IBE scheme in 2001.

Around the same time, another IBE scheme was published by Cocks [70]. This does not use pairings. In fact, it is particularly simple and does not use anything more than quadratic residues. Cocks works for the British government (Government Communications Headquarters (GCHQ)) and has been credited [164] for discovering the RSA system earlier to its appearance in the public domain. It is interesting that one of the public inventors of RSA scheme proposed the problem of constructing an IBE scheme which was solved by one of the government inventors of the same RSA system.

The IBE scheme for Cocks encrypts one bit at a time. The resulting ciphertext size is quite large (though it is still polynomial in the size of the security parameter). Boneh, Gentry and Hamburg [40] proposed another IBE scheme without pairings which has a more compact ciphertext. But, the trade-off is that it is more inefficient compared to Cocks IBE scheme. There have been some follow up work along this line [111], but, the potential of this line of research is not quite clear.

Another technique for constructing IBE schemes have arisen in the last few years. This is to use lattice-based methods. Lattices have been around for quite some time and have been used in cryptology for both cryptanalysis (as in Shamir's attack on knapsack PKE schemes [156]) and cryptography (including design of hash functions [7] and PKE schemes [97, 105, 147]). A paper by Gentry, Peikert and Vaikuntanathan [93] introduced a new technique called pre-image sampling and showed how to realise this using lattices. As a result, a whole collection of new constructions for several important cryptographic primitives such as signatures and IBE schemes was made possible. The theme of constructing lattice-based IBE schemes was followed upon by later authors who proposed progressively improved constructions.

The purpose of this chapter is to provide an idea of the non-pairing based IBE schemes. We discuss Cocks' IBE scheme and the basic idea of the Boneh-Gentry-Hamburg IBE scheme. For lattices, we describe the basic design strategy introduced by Peikert, Gentry and Vaikuntanathan and mention some of the later extensions that have been proposed. The security reductions are not presented. For these we refer the reader to the respective papers.

9.1 IBE Based on Number Theory

In this section, we discuss Cocks' method and the Boneh-Gentry-Hamburg method.

9.1.1 Cocks' IBE

An elegant solution to the problem of constructing an IBE was given by Cocks [70]. Briefly, the idea is the following. Our description is based on [40].

For a positive integer N, the set $QR(N)$ consists of all quadratic residues modulo N and $J(N)$ is the set of all elements of $\mathbb{Z}_N$ having Jacobi symbol 1. An element in $J(N) \setminus QR(N)$ is a non-square having Jacobi symbol 1.

Set-Up. The public parameters of the system consists of $N = pq$ (where p and q are primes such that factoring N is hard); a random u from $J(N) \setminus QR(N)$; and a hash function $H()$ which maps identities into $J(N)$. The master secret key is the factorization (p, q) of N.

KeyGen. The secret key corresponding to an identity id is obtained by first computing $R = H(\text{id})$ and then r to be either $\sqrt{R}$ or $\sqrt{uR}$ according as R is a square modulo N or not. The secret key for id is r.

Encrypt. To encrypt a bit m (represented as $+1$ or -1) using an identity id, compute R from id as above; randomly choose t_0, t_1 from $\mathbb{Z}_N$ and for $a \in \{0, 1\}$ compute $d_a = (t_a^2 + u^a R)/t_a$ and $c_a = m \cdot (\frac{t_a}{N})$. The ciphertext consists of the two elements $((d_0, c_0), (d_1, c_1))$.

Decrypt. For decryption using identity id and secret key r, first set $a \in \{0, 1\}$ such that $r^2 = u^a R$, where R is obtained from id as above. Set $g = d_a + 2r$. Note that $g = \left(\frac{(t_a + r)^2}{t_a} \right)$ and hence, $(\frac{g}{N}) = (\frac{t_a}{N})$. So, the receiver can compute $m = c_a \cdot (\frac{g}{N})$.

While this is an interesting protocol, one main problem with it is that the size of the ciphertext is very large. The overhead consists of two elements of $\mathbb{Z}_N$ per bit of the message.

The security of the system is based on the quadratic residuacity (QR) assumption under the random oracle model. The QR assumption is that given an element x of $J(N)$ it is hard to determine whether x is actually a square modulo N, or whether it is a pseudo-square, i.e., it is a non-square modulo both p and q. If N can be factored into its constituent primes, then it is easy to solve the problem of determining the quadratic character of x. So, the hardness of the quadratic residuacity problem presupposes the hardness of the factoring problem.

9.1.2 Boneh-Gentry-Hamburg IBE

Boneh, Gentry and Hamburg [40] have given a non-pairing based IBE which is more space efficient compared to Cocks method. The construction is described in two parts. In the first part, an IBE is described which encrypts a single bit. This is a general description of which the Cocks-IBE is *not* an instantiation. In the second part, they show how to reuse the random elements for more than one bit. It is such reuse which significantly reduces the size of the ciphertext compared to Cocks' scheme.

Below we describe the IBE from [40] for which the ciphertext length is $\lceil \log_2 N \rceil + 2\ell$ for an ℓ-bit message and $N = pq$ as before. This construction is called **BasicIBE** in [40]. Another construction is described for which the ciphertext is $\lceil \log_2 N \rceil + (\ell + 1)$ bits long. The description of this construction is omitted.

BasicIBE.

Let $\mathscr{I}\mathscr{D}$ denote the identity space.

Set-Up. Let $N = pq$, where p and q are primes such that it is difficult to factorize N. Choose a uniform random u from $J(N) \setminus QR(N)$. Let $H : \mathscr{I}\mathscr{D} \times \{1,\ldots,\ell\} \to J(N)$ be a hash function which the security reduction models as a random oracle. The public parameters consists of (N, u, H). The master secret key is (p, q) along with a key K for a pseudo-random function $F_K : \mathscr{I}\mathscr{D} \times \{1,\ldots,\ell\} \to \{0,1,2,3\}$.

KeyGen. The input consists of an identity id, the master secret key (p, q) and a message length parameter ℓ. The decryption key d_{id} corresponding to id is $(r_1,\ldots,r_\ell)$ where the r_j's are obtained as follows.

For $j = 1,\ldots,\ell$ do
  $R_j = H(\text{id}, j) \in J(N)$;
  $w = F_K(\text{id}, j) \in \{0, 1, 2, 3\}$;
  let $a \in \{0, 1\}$ be such that $u^a R_j \in QR(N)$;
  let $\{z_0, z_1, z_2, z_3\}$ be the four square roots of $u^a R_j$ in $\mathbb{Z}_N$;
  set $r_j = z_w$;
end.

Encrypt. The input consists of the public parameters (N, u, H), an identity id and a message $m = m_1 \cdots m_\ell \in \{-1, 1\}^\ell$. Generate a uniform random $s \in \mathbb{Z}_N$ and set $S = s^2$. The ciphertext is $(S, \mathbf{c}, \bar{\mathbf{c}})$ where $\mathbf{c} = (c_1,\ldots,c_\ell)$ and $\bar{\mathbf{c}} = (\bar{c}_1,\ldots,\bar{c}_\ell)$ and these are generated as follows.

For $j = 1, \ldots, \ell$ do
    $R_j = H(\text{id}, j)$;
    $(f, g) = \mathcal{Q}(N, R_j, S)$;
    $(\overline{f}, \overline{g}) = \mathcal{Q}(N, uR_j, S)$;
    $c_j = m_j \cdot \left(\frac{g(s)}{N} \right)$;
    $\overline{c}_j = m_j \cdot \left(\frac{\overline{g}(s)}{N} \right)$;
end.

Decrypt. The input is a ciphertext $(S, \mathbf{c}, \overline{\mathbf{c}})$ encrypted to an identity id and the corresponding decryption key $d_{\text{id}} = (r_1, \ldots, r_\ell)$.

For $j = 1, \ldots, \ell$ do
    $R_j = H(\text{id}, j)$;
    if $r_j^2 = R_j$
        $(f, g) = \mathcal{Q}(N, R_j, S)$;
        set $m_j = c_j \cdot \left(\frac{f(r_j)}{N} \right)$;
    if $r_j^2 = uR_j$
        $(\overline{f}, \overline{g}) = \mathcal{Q}(N, uR_j, S)$;
        set $m_j = \overline{c}_j \cdot \left(\frac{\overline{f}(r_j)}{N} \right)$;
end;
output $m_1 \ldots, m_\ell$.

The function H generates elements in $J(N)$. Given a function H' which generates elements in $\mathbb{Z}_N$, it is easy to construct H. Fix a publicly known element $z \in \mathbb{Z}_N$ whose Jacobi symbol is -1 and let $x = H'(\text{id}, j)$. Now return either x or zx as the output of $H(\text{id}, j)$ according as whether the Jacobi symbol of x is 1 or -1.

The algorithm $\mathcal{Q}$ in the above description is a deterministic algorithm which takes three inputs – a positive integer N and two elements of $\mathbb{Z}_N$ and returns two polynomials f, g satisfying the following conditions.

- If R and S are quadratic residues and r and s are respectively any square roots of R and S, then $f(r)g(s)$ is a quadratic residue.
- If R is a quadratic residue and r is any square root of R, then $f(r)f(-r)S$ is a quadratic residue.

From the first condition, it follows that $(f(r)/N)$ is equal to $(g(s)/N)$. The soundness of decryption follows from this condition. The second condition is required for the security reduction to go through. The following example of $\mathcal{Q}$ has been described in [40]: given input (N, R, S), construct a solution $(x, y) \in \mathbb{Z}_N^2$ to the equation $Rx^2 + Sy^2 = 1$ and output the polynomials $f(r) = xr + 1$ and $g(s) = 2ys + 2$. The main difficulty in the scheme is in finding a suitable solution (x, y). Several approaches are described in [40], but, none of them are at the level of practicality for the accepted security level. Another approach to this problem is discussed in [111].

9.2 IBE From Lattices

We start by mentioning some of the background on lattices. This description is the minimal required to get some understanding of the IBE constructions. Lattices have a rich mathematical structure and associated computational problems. The interested reader is referred to [142] for a more comprehensive background.

9.2.1 Background on Lattices

Lattices are defined to be discrete subgroups of $\mathbb{R}^m$, where m is a positive integer. More conveniently, lattices are defined as follows. An n-dimensional lattice of rank $k \leq n$ is

$$\Lambda = \mathscr{L}(\mathbf{B}) = \left\{ \mathbf{Bc} : \mathbf{c} \in \mathbb{Z}^k \right\}$$

where the k columns $\mathbf{b}_1, \ldots, \mathbf{b}_k \in \mathbb{R}^n$ of the basis $\mathbf{B}$ are linearly independent. The parameter n is taken to be the security parameter and all other parameters are usually considered to be functions of n.

The length of a vector $\mathbf{x}$ in a lattice is measured by its norm $||\mathbf{x}||$. The most common norm is the Euclidean norm defined as

$$||\mathbf{x}|| = \sqrt{\sum_{i=1}^{n} x_i^2}.$$

The minimum distance $\lambda_1(\Lambda)$ of a lattice Λ is defined to be $\min_{\mathbf{x} \neq \mathbf{y}} ||\mathbf{x} - \mathbf{y}||$, where $\mathbf{x}$ and $\mathbf{y}$ are elements of Λ. The minimum distance is equal to the length of its shortest non-zero element, i.e., $\lambda_1(\Lambda) = \min_{\mathbf{x} \in \Lambda, \mathbf{x} \neq \mathbf{0}} ||\mathbf{x}||$. Given a set S of lattice points, $||S||$ is defined to be $\max ||\mathbf{s}||$ where the maximum is taken over all elements $\mathbf{s} \in S$.

Let q be a prime. (*In the previous chapters, q denoted the number of queries made by an adversary, but, in the literature on lattices it is more customary to denote the underlying prime by q.*) Let $\mathbf{A} \in \mathbb{Z}_q^{n \times m}$, i.e., $\mathbf{A}$ is an $n \times m$ matrix with entries from $\mathbb{Z}_q$. Define the following two kinds of lattices.

$\Lambda(\mathbf{A}, q) = \{ \mathbf{y} \in \mathbb{Z}^m : \mathbf{y} = \mathbf{A}^T \mathbf{s} \bmod q \text{ for some } \mathbf{s} \in \mathbb{Z}^n \}$.
$\Lambda^{\perp}(\mathbf{A}, q) = \{ \mathbf{e} \in \mathbb{Z}^m : \mathbf{Ae} = \mathbf{0} \bmod q \}$.

Here $\mathbf{A}^T$ is the transpose of $\mathbf{A}$. In terms of coding theory, $\mathbf{A}^T$ is the generator matrix for the lattice $\Lambda(\mathbf{A}, q)$ and $\mathbf{A}$ is the parity check matrix for the lattice $\Lambda^{\perp}(\mathbf{A}, q)$. Lattices defined over $\mathbb{Z}_q$ are called modular lattices.

There are two classical hard problems on lattices: the shortest vector problem (SVP) and the closest vector problem (CVP). An instance of the decision version of the shortest vector problem $\mathsf{GapSVP}_{\gamma(n)}$ is a pair $(\mathbf{B}, d)$ where $\mathbf{B}$ is a full-rank n-dimensional lattice and $d \in \mathbb{R}$. It is a 'Yes' instance if $\lambda_1(\mathscr{L}(\mathbf{B})) \leq d$ and it is a 'No' instance if $\lambda_1(\mathscr{L}(\mathbf{B})) > \gamma(n)d$. The closest vector problem is defined similarly.

A related problem $\mathsf{SIVP}_{\gamma(n)}$ takes as input a full-rank basis $\mathbf{B}$ of an n-dimensional lattice and has to produce a set of n linearly independent lattice vectors $\mathbf{S} \subset \mathscr{L}(\mathbf{B})$ such that $||\mathbf{S}|| \leq \gamma(n)\lambda_n(\mathscr{L}(\mathbf{B}))$, where $\lambda_i(\Lambda)$ (called the i-th successive minimum) is the smallest radius r such that Λ contains i linearly independent vectors of norm at most r.

The learning with errors (LWE) problem was introduced by Regev [147] and is defined as follows. Fix q and let χ be a probability distribution on $\mathbb{Z}_q$. An instance of $\mathsf{LWE}_{q,\chi}$ is a pair $(\mathbf{a}, v)$, where $\mathbf{a} \in \mathbb{Z}_q^n$ is chosen uniformly at random and v is either chosen uniformly at random from $\mathbb{Z}_q$ or is chosen to be $\mathbf{a}^T\mathbf{s} + x$, where $\mathbf{s}$ is in $\mathbb{Z}_q^n$ and x is chosen from $\mathbb{Z}_q$ according to the distribution χ. The goal is to distinguish between these two cases.

Let $\mathbf{T} = \mathbb{R}/\mathbb{Z}$ be the group of reals $[0, 1)$ with modulo 1 addition. For a positive real α, let Ψ_α be the distribution on $\mathbf{T}$ of a normal variable with mean 0 and standard deviation $\alpha/\sqrt{2\pi}$ reduced modulo 1. Given a positive integer q (which is often implicitly given and is not necessarily prime), let the discretization $\overline{\Psi_\alpha}$ be the discrete distribution over $\mathbb{Z}_q$ of the random variable $\lfloor q \cdot X \rceil$ where X has the distribution Ψ_α. Here $\lfloor x \rceil$ is the nearest integer to x and is defined to be $\lfloor x \rceil = \lfloor x + 1/2 \rfloor$.

The importance of the LWE problem arises from a result by Regev [147] which showed that for $\alpha = \alpha(n) \in (0, 1)$ and $q = q(n)$ (i.e., α and q are functions of a security parameter n) if there exists an efficient (possibly quantum) algorithm to solve $\mathsf{LWE}_{q,\overline{\Psi_\alpha}}$, then there exists an efficient quantum algorithm for approximating SIVP and GapSVP in the Euclidean norm, in the worst case, to within $\tilde{O}(n/\alpha)$ factors.

Define the Gaussian function on $\mathbb{R}^n$ centred at a point $\mathbf{c} \in \mathbb{R}^n$ with parameter $s \in \mathbb{R}$ to be

$$\rho_{s,\mathbf{c}}(\mathbf{x}) = \exp\left(\frac{-\pi||\mathbf{x} - \mathbf{c}||}{s^2}\right), \quad \text{for all} \quad \mathbf{x} \in \mathbb{R}^n.$$

The total measure associated to $\rho_{s,\mathbf{c}}$ is $\int_{\mathbf{x} \in \mathbb{R}^n} \rho_{s,c}(\mathbf{x})d\mathbf{x} = s^n$. Scaling by s^n, it is possible to define the continuous Gaussian distribution around $\mathbf{c}$ having density function $D_{s,\mathbf{c}}(\mathbf{x})$ where for all $\mathbf{x} \in \mathbb{R}^n$,

$$D_{s,\mathbf{c}}(\mathbf{x}) = \frac{\rho_{s,\mathbf{c}}(\mathbf{x})}{s^n}.$$

The Gaussian ρ extends straightforwardly to a countable set of points A as

$$\rho_{s,\mathbf{c}}(A) = \sum_{\mathbf{x} \in A} \rho_{s,\mathbf{c}}(\mathbf{x}).$$

$D_{s,\mathbf{c}}$ can be expressed as the sum of n orthogonal 1-dimensional Gaussian distributions. Each of these 1-dimensional distributions can be efficiently approximated with arbitrary precision. So, $D_{s,\mathbf{c}}$ can be efficiently approximated. It is more convenient for our understanding to work with real numbers while keeping in mind that in practice finite precision can be used to choose points with probability approximately proportional to $D_{s,\mathbf{c}}$.

Given an n-dimensional lattice Λ of rank $k \leq n$, define the discrete Gaussian distribution over Λ as follows.

$$D_{\Lambda,s,\mathbf{c}} = \frac{\rho_{s,\mathbf{c}}(\mathbf{x})}{\rho_{s,\mathbf{c}}(\Lambda)} \quad \text{for all} \quad \mathbf{x} \in \mathbb{R}^n.$$

The distribution $D_{\Lambda,s,\mathbf{c}}$ is called a discrete Gaussian distribution. It is possible to view $D_{\Lambda,s,\mathbf{c}}$ as a conditional distribution, i.e., $D_{\Lambda,s,\mathbf{c}}(\mathbf{x})$ is the probability of $\mathbf{x}$ conditioned on the event that $\mathbf{x} \in \Lambda$. This can be seen by considering that if $\mathbf{x}$ is chosen from a very fine grid and α is the volume of one cell, then, the probability of $\mathbf{x}$ according to $D_{s,\mathbf{c}}$ is $\alpha D_{s,\mathbf{c}}(\mathbf{x})$ while the probability that $\mathbf{x}$ is in Λ is $\alpha D_{s,\mathbf{c}}(\Lambda)$.

Gentry, Peikert and Vaikuntanathan [93] showed a method to sample a lattice point following a discrete Gaussian distribution. The method they introduce is different from an earlier method by Regev [147]. Details of the sampling method are a bit involved as is the proof of its correctness. Here we will not get into these details, but, only record the following fact.

Fact 1. There is an algorithm SampleD, which for any lattice basis $\mathbf{B} \in \mathbb{Z}^{n \times k}$, any real $s \geq \|\mathbf{B}\| \cdot \omega\left(\sqrt{\log n}\right)$ and any $\mathbf{c} \in \mathbb{R}^n$, produces an output SampleD$(\mathbf{B}, s, \mathbf{c})$ the distribution of which is within negligible statistical distance from $D_{\mathscr{L}(\mathbf{B}),s,\mathbf{c}}$. The run-time of SampleD is polynomial in n and the size of its input.

The small integer solution (SIS) problem in the Euclidean norm is the following. Given an integer q, a matrix $\mathbf{A} \in \mathbb{Z}_q^{n \times m}$ and a real β, find a non-zero integer vector $\mathbf{e} \in \mathbb{Z}^m$ such that $\mathbf{Ae} = \mathbf{0} \bmod q$ and $\|\mathbf{e}\| \leq \beta$. This problem asks for a pre-image of $\mathbf{0}$. A natural variant called the inhomogeneous small integer solution (ISIS) problem specifies a vector $\mathbf{u} \in \mathbb{Z}_q^n$ and asks for a pre-image of $\mathbf{u}$, i.e., an $\mathbf{e}$ such that $\mathbf{Ae} = \mathbf{u} \bmod q$. The other conditions remain the same as for the SIS problem.

9.2.2 Pre-Image Sampling

A collection of trapdoor functions with pre-image sampling is a collection of three algorithms (TrapGen, SampleDom, SamplePre) such that given a security parameter n, TrapGen(1^n) outputs a pair (a, t) where a is the description of an efficiently computable function $f_a : D_n \rightarrow R_n$ and t is some trapdoor information for f_a. The algorithm SampleDom(1^n) samples an x from some distribution over D_n such that the distribution of $f_a(x)$ is uniform over R_n for every function description a. Algorithm SamplePre(t, y) allows the sampling of an element from the set of all pre-images of y, i.e., it returns an x following SampleDom conditioned on the event $f_a(x) = y$. There is an additional condition that without the trapdoor, f_a behaves as a one-way function. A modification can be used to define a collection of trapdoor collision-resistant hash functions with pre-image sampling. See [93] for details.

Gentry, Peikert and Vaikuntanathan [93] show that it is possible to build a collection of trapdoor functions with pre-image sampling using hard problems from lattices. Their core idea is the sampling algorithm SampleD which shows how to sample from a lattice following a distribution which is statistically close to a discrete Gaussian distribution. The construction is given by the following three points.

1. The first step is to show a trapdoor generator. This follows from an early result of Ajtai [8], which states that for any prime q (polynomially bounded in the security parameter n) and any $m \geq 5n\log_2 q$, there is a probabilistic polynomial time algorithm that on input 1^n outputs a matrix $\mathbf{A} \in \mathbb{Z}_q^{n \times m}$ and a full-rank set $\mathbf{S} \subset \Lambda^\perp(\mathbf{A}, q)$ where the distribution of A is statistically close to uniform over $\mathbb{Z}_q^{n \times m}$ and $||\mathbf{S}|| \leq L = m^{2.5}$. With overwhelming probability $\mathbf{A}$ is of rank n.
2. The function $f_\mathbf{A}$ is defined as $f_\mathbf{A}(\mathbf{e}) = \mathbf{A}\mathbf{e} \bmod q$ with domain $D_n = \{\mathbf{e} \in \mathbb{Z}^m : ||\mathbf{e}|| \leq s\sqrt{m}\}$ and range $R_n = \mathbb{Z}_q^n$. The distribution on D_n is $D_{\mathbb{Z}^m, s}$ which can be sampled using SampleD with the standard basis for $\mathbb{Z}^m$.
3. The trapdoor inversion algorithm is SampleISIS$(\mathbf{A}, \mathbf{S}, s, \mathbf{u})$ which samples from $f_\mathbf{A}^{-1}(\mathbf{u})$ as follows. First use standard linear algebraic techniques to obtain a $\mathbf{t} \in \mathbb{Z}^m$ such that $\mathbf{A}\mathbf{t} = \mathbf{u} \bmod q$. Then sample $\mathbf{v}$ from the distribution $D_{\Lambda^\perp, s, -\mathbf{t}}$ using the SampleD algorithm with the trapdoor $\mathbf{S}$ and output $\mathbf{e} = \mathbf{t} + \mathbf{v}$.

It is shown in [93] that the above provides a collection of trapdoor one-way functions with pre-image sampling under the assumption that the ISIS problem is hard. Further, the one-wayness property can be replaced by collision-resistance assuming that the SIS problem is hard.

9.2.3 Gentry-Peikert-Vaikuntanathan IBE

It has been mentioned earlier that following an observation of Naor, an IBE scheme secure against adaptive-identity attacks gives rise to a signature scheme which is existentially unforgeable against chosen message attacks. Gentry in [91] had suggested that following this observation, to construct an IBE scheme one starts out in some way by designing a signature scheme.

A collection of trapdoor collision-resistant functions with pre-image sampling gives rise to a very natural signature scheme [93]. The key generation algorithm outputs a pair (a, t) where a is the description of a public efficiently computable function f_a and t is a trapdoor for f_a. The signing key is t while the verification key is a. Given a message M, it is hashed into the range using a hash function H (modelled as a random oracle) and the signature σ is simply a pre-image of $H(M)$ under f_a and computed using t. To verify, one simply computes $f_a(\sigma)$ and checks whether this is equal to $H(M)$. This simple idea can be shown to yield a signature scheme which is strongly unforgeable under chosen message attacks.

The idea of obtaining a signature on a message is to obtain a pre-image of the message hashed into the range. In the context of IBE, a decryption key for an identity can be seen as the signature of the PKG on the identity. So, by hashing the identity into the range it is possible to obtain a decryption key by generating one of the pre-images. Though conceptually simple, there are a few subtleties involved in ensuring that a random oracle can map identities into the range set. The details for a single-bit message are given below.

Denote by n the security parameter. Let $q = q(n)$ be a prime and $m = m(n)$ (typically $O(n \log n)$) be a positive integer which will be the dimension of the lattice.

Set-Up. Using the previously mentioned trapdoor generator, generate a matrix $\mathbf{A} \in \mathbb{Z}_q^{n \times m}$ and a trapdoor $\mathbf{S} \subseteq \Lambda^{\perp}(\mathbf{A}, q)$ such that the following holds: for any $\mathbf{u} \in \mathbb{Z}_q^n$, using the trapdoor $\mathbf{S}$, it is possible to sample an $\mathbf{e} \in \mathbb{Z}_q^m$ from the set of all pre-images of $\mathbf{u}$ under f.

The public parameter for the IBE scheme consists of the matrix $\mathbf{A}$ and the master secret key is $\mathbf{S}$.

Key-Gen. A decryption key generated for an identity is stored and if there is a future request for this identity, then the previously generated key is returned.

The definition of the decryption key for an identity $\text{id} \in \{0,1\}^*$ requires a hash function $H : \{0,1\}^* \to \mathbb{Z}_q^n$ (which is modelled in the proof as a random oracle). Given id, let $\mathbf{u} = H(\text{id})$ and the master secret $\mathbf{S}$ is used to obtain a pre-image $\mathbf{e}$ of $\mathbf{u}$ under f. The decryption key for id is defined to be $\mathbf{e}$.

Let χ be a distribution on $\mathbb{Z}_q$ and χ^m is the m-fold product distribution on $\mathbb{Z}_q^m$. The distribution χ is parameterized by a parameter r which is chosen so as to ensure that the difficulty of breaking the resulting IBE scheme can be based on the difficulty of the "learning with errors" (LWE) problem for the underlying lattice.

Encrypt. Encryption of a bit b to an identity id is done as follows: let $\mathbf{u} = H(\text{id}) \in \mathbb{Z}_q^n$; choose $\mathbf{s}$ uniformly from $\mathbb{Z}_q^n$ and set $\mathbf{p} = \mathbf{A}^T\mathbf{s} + \mathbf{x}$ where $\mathbf{x}$ is drawn from $\mathbb{Z}_q^m$ according to the distribution χ^m. The ciphertext consists of the pair $(\mathbf{p}, c)$ where $c = \mathbf{u}^T\mathbf{s} + x + b \cdot \lfloor q/2 \rfloor$ and x is chosen from $\mathbb{Z}_q$ according to χ.

Decrypt. Given a ciphertext $(\mathbf{p}, c)$, an identity id and a corresponding decryption key $\mathbf{e}$, the recovery of the encrypted bit proceeds as follows: compute $b' = c - \mathbf{e}^T\mathbf{p} \in \mathbb{Z}_q$ and output 0 if b' is closer to 0 than to $\lfloor q/2 \rfloor$ modulo q, otherwise output 1.

The encryption algorithm samples a vector $\mathbf{p} = \mathbf{A}^T\mathbf{s} + \mathbf{x}$ for uniform random $\mathbf{s}$ and $\mathbf{x}$ chosen from χ^m. This vector is essentially an LWE vector. The vector $\mathbf{u}$ is used to generate another LWE instance $p = \mathbf{u}^T\mathbf{s} + x$ where x is chosen from χ. This value p is used to mask the message b (transformed as $b \cdot \lfloor q/2 \rfloor$). Since $\mathbf{u}$ is uniform the adversary's view is indistinguishable from uniform (under the assumption that LWE is hard). For this reason, the scheme is also anonymous, i.e., given a ciphertext it is not possible to decide the identity to which it was encrypted.

More formally, it can be shown that for a properly encrypted message, decryption is successful with an overwhelming probability and that the scheme is CPA-secure and anonymous assuming that a suitably parameterized version of the LWE problem is hard.

The encryption algorithm proceeds one bit at a time. As a result, the size of the ciphertext is quite large. To a certain extent this problem can be tackled in the following manner. Suppose that the message is a k-bit string. Identities are mapped by H to k elements $\mathbf{u}_1,\ldots,\mathbf{u}_k \in \mathbb{Z}_q^n$ where $\mathbf{u}_i$ is used to encrypt the i-th bit of the message. The randomness $\mathbf{s}$ is used for all the bits and so the vector $\mathbf{p} = \mathbf{A}^T\mathbf{s} + \mathbf{x}$ remains the same for all the encryptions. Note that as a trade-off there will be k elements of $\mathbb{Z}_q^m$ as part of the secret key.

9.2.4 Generalized Pre-Image Sampling

Subsequent works have provided different constructions of (H)IBE both in the random oracle model and without it. A key idea in the above construction is that of pre-image sampling. Non-trivial extensions of the construction are made by developing on the idea of pre-image sampling. More general forms of pre-image sampling and constructions of both IBE and HIBE schemes have been given in [56, 5, 6].

Generalised pre-image sampling has been introduced in [55]. At a broad level, the idea is the following. Given a pair (a,t) where a describes a function f_a and t is a trapdoor for the function, for any y it is shown how to sample from the set of pre-images $f_{a'}^{-1}(y)$, where a' is somehow related to a. The relationship in the context of lattices is that a' is given by a block matrix $\mathbf{A} = [\mathbf{A}_1,\ldots,\mathbf{A}_k]$ where each $\mathbf{A}_i \in \mathbb{Z}_q^{n \times m}$ and a is given by a block matrix $\mathbf{A}_T = [\mathbf{A}_{i_1},\ldots,\mathbf{A}_{i_j}]$ for some $T = \{i_1,\ldots,i_j\} \subseteq \{1,\ldots,k\}$.

The strategy for the more general version of pre-image sampling in [55] is simple. Suppose $\mathbf{A} = [\mathbf{A}_1,\mathbf{A}_2]$ and that we have a short basis $\mathbf{S}_1$ for $\Lambda^{\perp}(\mathbf{A}_1)$. Given a $\mathbf{y} \in \mathbb{Z}_q^n$, the goal is to sample from the set of pre-images of $\mathbf{y}$ under $\mathbf{A}$ using the short basis $\mathbf{S}_1$. As a first step, sample $\mathbf{e}_2$ from $D_{\mathbb{Z}^m,s}$ and define $\mathbf{z} = \mathbf{y} - \mathbf{A}_2\mathbf{e}_2$. Now, if we can sample from the set of pre-images of $\mathbf{z}$ under $\mathbf{A}_1$, then we are done since $(\mathbf{e}_1^T,\mathbf{e}_2^T)^T$ is a required pre-image. This is made possible by the short basis $\mathbf{S}_1$ of $\Lambda^{\perp}(\mathbf{A}_1)$ that is given. The key step is to change the image point $\mathbf{y}$ for the entire matrix $\mathbf{A}$ to the image point $\mathbf{z}$ for the sub-matrix $\mathbf{A}_1$. Clearly, this strategy works for the more general case described in the earlier paragraph. The technical difficulty in proving the correctness of this method is to show that the sampled pre-image has the required distribution.

This idea extends to finding a short basis for $\mathbf{A}$ given a basis for $\mathbf{A}_1$ (or $\mathbf{A}_T$ in the more general case). The idea is to repeatedly sample from the set of pre-images of $\mathbf{y} = \mathbf{0}$ for $\mathbf{A}$ using the short basis $\mathbf{S}_1$ of $\Lambda^{\perp}(\mathbf{A}_1)$. Having sufficiently many samples ensures that with high probability, there are $2m$ (or km in the general case) linearly independent vectors in the sampled set. Now a standard result in lattices can be used to ensure that it is possible to pick $2m$ sufficiently short linearly independent vectors giving a basis for $\Lambda^{\perp}(\mathbf{A})$.

A net result is that given a short basis for $\Lambda^{\perp}(\mathbf{A}_1)$ one can obtain a short basis for $\Lambda^{\perp}(\mathbf{A})$. Clearly, this strategy can be extended further, i.e., given a short basis for $\Lambda^{\perp}([\mathbf{A}_1,\ldots,\mathbf{A}_{i-1}])$, it is possible to obtain a short basis for $\Lambda^{\perp}([\mathbf{A}_1,\ldots,\mathbf{A}_{i-1},\mathbf{A}_i])$. (The quality of the basis degrades but, is still good enough for pre-image sampling.) This gives rise to the idea of delegating a basis and hence obtaining a HIBE scheme. Consider $\mathbf{A}_i$ to be the portion of the public parameter associated with the

i-the level of the HIBE. A decryption key for the i-th level is a short basis for $\Lambda^{\perp}([\mathbf{A}_1,\ldots,\mathbf{A}_{i-1},\mathbf{A}_i])$. Now the basis delegation technique outlined above ensures that given a decryption key for an $(i-1)$-level identity it is possible to obtain a decryption key for an i-level identity. Combining this idea with the encryption method of the Gentry-Peikert-Vaikuntanathan IBE scheme gives an HIBE scheme much like the way the Boneh-Franklin IBE scheme gives rise to the Gentry-Silverburg HIBE scheme. Security of all these schemes are obtained under the random oracle assumption.

9.2.5 Agrawal-Boneh-Boyen IBE

A use of the general pre-image sampling technique given in [5] is to obtain an IBE scheme which is secure without random oracles. Along with the pre-image sampling method some additional intuition is required. In the Gentry-Peikert-Vaikuntanathan IBE scheme of the previous section, an identity is mapped to an element $\mathbf{u}$ and the trapdoor $\mathbf{S}$ is used to find a pre-image. The matrix $\mathbf{A}$ defines the forward computation and does not depend on the identity. At a broad level, this can be thought to be similar to the Boneh-Franklin IBE scheme where identities are mapped into elliptic curve points and the decryption key is the s-fold scalar multiple of this point.

A different way to approach the problem of constructing a lattice-based IBE scheme while keeping within the framework of pre-image sampleable trapdoor function is to change the roles of the range point and the matrix mapping. In [5], the range point $\mathbf{u}$ is fixed and independent of the identity. The matrix $\mathbf{A}$ is changed to $[\mathbf{A}_0|\mathbf{A}_1 + H(\mathsf{id})\mathbf{B}]$, where the matrices $\mathbf{A}_0, \mathbf{A}_1$ and $\mathbf{B}$ are specified as part of the public parameters and $H : \mathbb{Z}_q^n \to \mathbb{Z}_q^{n \times n}$ is a public function which maps an identity (considered to be an element of $\mathbb{Z}_q^n$) to a matrix. The function H maps $\mathbb{Z}_q^n$ to $\mathbb{Z}_q^{n \times n}$ and satisfies a property called "full rank difference" in [5]: for any two distinct $\mathbf{u}, \mathbf{v} \in \mathbb{Z}_q^n$, the matrix $H(\mathbf{u}) - H(\mathbf{v})$ is full rank.

This change of roles is to be seen in conjunction with the general pre-image sampling method of [56]. Consider a matrix $\mathbf{F} = [\mathbf{A}, \mathbf{AR} + \mathbf{B}]$, where $\mathbf{A}$ and $\mathbf{B}$ are in $\mathbb{Z}_q^{n \times m}$ and $\mathbf{R} \in \{-1, 1\}^{m \times m}$. Two pre-image sampling methods SampleLeft and SampleRight are described. SampleLeft works with a short basis for $\mathbf{A}$ and is based on the generalised pre-image sampling in [5] described above. SampleRight is based on a lattice delegation technique from [146]. In the actual IBE scheme, only SampleLeft is required whereas SampleRight is required during the simulation in the security argument. The details of the Agrawal-Boneh-Boyen IBE scheme is as follows.

Set-Up. Given n and q, use the usual trapdoor generation method (due to Ajtai [8]) to generate a matrix $\mathbf{A}_0 \in \mathbb{Z}_q^{n \times m}$ and a short basis $\mathbf{S}_0$ for $\Lambda^{\perp}(\mathbf{A}_0)$. Select two uniform random matrices $\mathbf{A}_1$ and $\mathbf{B}$ from $\mathbb{Z}_q^{n \times m}$. Select a uniform random $\mathbf{u} \in \mathbb{Z}_q^n$. The public parameters consist of $(\mathbf{A}_0, \mathbf{A}_1, \mathbf{B}, \mathbf{u})$ whereas the master secret key is $\mathbf{S}_0$.

Identities are considered to be elements of $\mathbb{Z}_q^n$. For any identity id, let

$$F_{id} = [A_0, A_1 + H(id)B].$$

Key-Gen. Let id be an identity. Use the short basis S_0 of $\Lambda^\perp(A_0)$ to obtain a pre-image $e_{id} \in \mathbb{Z}_q^{2m}$ of u for F_{id}. (This is the SampleLeft algorithm which is the generalised pre-image sampling method from [55, 146].) Then $F_{id}e_{id} = u$. The pre-image e_{id} is a decryption key for the identity id.

Encrypt. Encryption of a bit b to an identity id is done as follows. Choose a uniform random $s \in \mathbb{Z}_q^n$ and a uniform random matrix $R \in \{-1, 1\}^{m \times m}$. Sample noise elements $x \in \mathbb{Z}_q$ and $y \in \mathbb{Z}_q^m$ respectively using distributions $\overline{\Psi}_\alpha$ and $\overline{\Psi}_\alpha^m$ and set $z = R^T y$. The ciphertext is (c_0, c_1) where

$$c_0 = u^T s + x + b\lfloor q/2 \rfloor, \quad c_1 = F_{id}^T s + \begin{bmatrix} y \\ z \end{bmatrix} \in \mathbb{Z}_q^{2m}.$$

Decrypt. Given a ciphertext (c_0, c_1) encrypted to an identity id and a decryption key e_{id} for id, decryption is done as follows. Compute $w = c_0 - e_{id}^T c_1 \in \mathbb{Z}_q$. Compare w and $\lfloor q/2 \rfloor$ as integers. If $|w - \lfloor q/2 \rfloor| < \lfloor q/4 \rfloor$, then output 1, else output 0.

Note that the masking of a message is similar to that of the previous scheme. A standard analysis shows that decryption succeeds with overwhelming probability. The matrix R plays a crucial role in the security reduction. Security is proved in the selective-identity model. The adversary specifies a target identity id* and then the simulator sets up the IBE scheme.

Security is based on the LWE problem. In the actual IBE scheme, a trapdoor for the matrix A_0 is known. On the other hand, in the instance of the LWE problem, the trapdoor for the matrix A_0 will not be known. But, the simulator in the proof still needs to be able to answer key extraction queries. This is done by generating a trapdoor for the matrix B and using the SampleRight algorithm as follows.

Suppose the adversary specifies id* as the target identity. The simulator now sets up the IBE scheme by first generating u in the usual manner. The simulator then generates a random matrix $A_0 \in \mathbb{Z}_q^{n \times m}$ and a pair (B, T) using the trapdoor generation algorithm, where $B \in \mathbb{Z}_q^{n \times m}$ and T is a short basis for the lattice $\Lambda^\perp(B)$. Generation of the challenge ciphertext will require a random matrix $R^* \in \{-1, 1\}^{m \times m}$. Since this does not depend on the adversary's queries, it is chosen during set-up. The matrix A_1 is defined to be

$$A_1 = A_0 R^* - H(id^*)B.$$

The public parameters (A_0, A_1, B, u) are declared. Note that the simulator does not possess a trapdoor for A_0 but does possess a trapdoor for B.

Suppose the adversary makes a key extraction query on an identity id. Then F_{id} is of the form

$$F_{id} = [A_0, A_1 + H(id)B] = [A_0, A_0 R^* + (H(id) - H(id^*))B].$$

By the full difference rank property of H, it follows that $H(\text{id}) - H(\text{id}^*)$ is non-singular. Also, with high probability $\mathbf{B}$ is non-singular and hence the matrix $\mathbf{B}' = (H(\text{id}) - H(\text{id}^*))\mathbf{B}$ is also non-singular. Knowledge of the trapdoor $\mathbf{T}$ for $\mathbf{B}$ allows the simulator to use the SampleRight algorithm to obtain a pre-image of $\mathbf{u}$ for $\mathbf{F}_{\text{id}}$. (Note that SampleRight requires a short basis for $\mathbf{B}'$ whereas the simulator actually has a short basis for $\mathbf{B}$. This fact is glossed over in [5].) So, the simulator is able to answer any key extraction query except on id^*.

The distribution of $(c_0^*, \mathbf{c}_1^*)$ in the actual scheme is that of an LWE instance where the input is from the "real" distribution. As is usual, in the last game of the security reduction this is taken from a "random" distribution, i.e., $(c_0^*, \mathbf{c}_1^*)$ is chosen uniformly at random from $\mathbb{Z}_q \times \mathbb{Z}_q^{2m}$. Then the adversary has no advantage in winning the selective-identity game. Also, it is shown that distinguishing between the two games is bounded above by the advantage of solving the LWE problem.

Extension to HIBE. Using the generalised pre-image sampling method of [55] described earlier or the delegation method of [146], it is possible to obtain a selective identity secure HIBE scheme. The essential idea is to be able to generate a trapdoor for a larger matrix from a trapdoor for a smaller matrix. To get a HIBE scheme, for an identity $\text{id} = (\text{id}_1, \ldots, \text{id}_\ell)$, the definition of $\mathbf{F}_{\text{id}}$ is changed to the following.

$$\mathbf{F}_{\text{id}} = [\mathbf{A}_0, \mathbf{A}_1 + H(\text{id}_1)\mathbf{B}, \ldots, \mathbf{A}_\ell + H(\text{id}_\ell)\mathbf{B}].$$

The matrices $\mathbf{A}_0, \mathbf{A}_1, \ldots, \mathbf{A}_\ell, \mathbf{B}$ are given in the public parameters. A decryption key for id is a pre-image of $\mathbf{u}$ for $\mathbf{F}_{\text{id}}$. This key, however, is not sufficient information to enable further delegation. For that a trapdoor is required. A basis delegation technique will allow the following. Consider an identity with i levels $(\text{id}_1, \ldots, \text{id}_i)$ and let $\text{ID}_{|i-1}$ be the identity consisting of the first $(i-1)$ components of id. Then $\mathbf{F}_{\text{id}_{|i-1}} = [\mathbf{A}_0, \mathbf{A}_1 + H(\text{id}_1)\mathbf{B}, \ldots, \mathbf{A}_{i-1} + H(\text{id}_{i-1})\mathbf{B}]$ and suppose that a short basis $S_{\text{id}_{|i-1}}$ is known for $\Lambda^\perp(\mathbf{F}_{\text{id}_{|i-1}})$. Then using generalised pre-image sampling (or the technique from [146]) it is possible to generate a short basis S_{id_i} for $\Lambda^\perp(\mathbf{F}_{\text{id}_i})$ where $\mathbf{F}_{\text{id}_i} = [\mathbf{A}_0, \mathbf{A}_1 + H(\text{id}_1)\mathbf{B}, \ldots, \mathbf{A}_i + H(\text{id}_i)\mathbf{B}]$.

This method of basis delegation increases the dimension of the lattice with increase in the length of the identity tuple. In [6], a HIBE construction is described where key delegation can be done without increasing lattice dimension.

Extension to adaptive-identity security. One may note that the simulation technique used in the security reduction is reminiscent of the simulation technique used for the Boneh-Boyen selective-identity secure IBE scheme (BB-IBE1). This scheme was modified by Waters [169] to obtain an adaptive-identity secure IBE scheme. In a similar manner, it is possible to modify the current scheme. Consider the identity as a sequence of ℓ bits $\text{id} = (\text{id}_1, \ldots, \text{id}_\ell) \in \{-1, 1\}^\ell$. The public parameters consist of matrices $\mathbf{A}_0$, $\mathbf{C}$ and matrices $\mathbf{A}_1, \ldots, \mathbf{A}_\ell$ one for each bit of the identity. The function $\mathbf{F}_{\text{id}}$ is now changed to the following.

$$\mathbf{F}_{\text{id}} = \left[\mathbf{A}_0, \mathbf{C} + \sum_{i=1}^{\ell} \text{id}_i \mathbf{A}_i\right].$$

The resulting IBE scheme will be adaptive-identity secure with shorter ciphertexts compared to the IBE scheme in [56]. The public key size is quite large, an $n \times m$ matrix for each bit of the identity. It is conceivable that the strategy used by Chatterjee-Sarkar [60] and Naccache [137] to reduce the public parameter size in Waters [169] scheme will also apply in the current context. This issue, however, is not addressed in [5].

Another issue not addressed in [5] is the question of adaptive-identity secure HIBE scheme. As suggested in Waters [169], using independent sets of public parameters for each level of the HIBE extends the IBE scheme in [169] to a HIBE scheme. Using completely independent public parameters for each level of the HIBE leads to a very large size for the public parameters. A modification suggested by Chatterjee and Sarkar [61] (as discussed in Chapter 7) is to reuse most of the public parameters for different levels and only introduce a single random element for each HIBE level. It is conceivable that this method will work in the current context. The details though need to be carefully worked out.

9.3 Conclusion

There are now two methods for realising IBE schemes – from bilinear pairings on elliptic curve groups and from lattices. The third approach is to use elementary number theoretic method as in Cocks and the Boneh-Gentry-Hamburg IBE schemes. Currently, this approach does not appear to be able to provide schemes which can compete in efficiency to schemes obtained from the first two approaches.

The motivation of lattices consists of three parts. These are relevant to any lattice-based cryptographic primitive and not particular to IBE schemes.

1. By an appropriate choice of parameters, the value of the prime q can be chosen small enough to fit in a machine sized word. This implies that implementation of lattice algorithms do not require efficient libraries for multi-precision arithmetic.
2. Security can be shown to be based on *worst* case instance of lattice problems.
3. Till date there are no quantum algorithms for solving hard computational problems for lattices.

On the other hand, the main problem for lattices based IBE schemes is the size of the keys and the ciphertexts. Compared to pairing based schemes these are far too large. We do not know whether any comparative study has been made of the speed of actual implementations of lattice based IBE schemes and pairing based IBE schemes. The requirement of multi-precision arithmetic library cannot by itself be considered a disadvantage. It is possible to develop good such libraries and in fact currently there are quite a few such implementations. Lastly, the issue of quantum algorithms not been known for lattice problems seems at best to be of theoretical interest. From a practical point of view, the construction of quantum computers do not seem to be anywhere on the horizon. So, pairing-based IBE schemes will continue to be of practical interest in the foreseeable future. At the same time, it is conceivable that researchers will pursue theoretical as well as practical works to come up with efficient and practical IBE schemes based on lattices.

Chapter 10
Applications, Extensions and Related Primitives

Techniques used to construct IBE schemes have been found to be useful in constructing other primitives. Some of these primitives can be seen as more general notions of IBE. The purpose of this chapter is to provide a brief discussion of these issues. This discussion is not meant to be comprehensive. Rather it is meant to provide the reader with an idea of the technical flexibility that the notion of (H)IBE offers.

We primarily discuss signature schemes, identity-based key agreement and broadcast encryption. Extensions such as fuzzy identity-based encryption and attribute-based encryption are only touched upon. These primitives have their own formal definitions and security models. We mostly describe these notions at an informal and intuitive level where we try to explain the connection to the techniques employed for (H)IBE schemes.

10.1 Signature Schemes

Signature schemes are among the most important cryptographic primitives. There is a signer who possesses a secret signing key and can sign messages using a signing algorithm. Using a matching public verification key, the message-signature pair can be checked for correctness using a verification algorithm.

10.1.1 Boneh-Lynn-Shacham Short Signature

An IBE scheme can be *generically* converted to a signature scheme as described below. For the conversion, the first thing to observe is that a decryption key corresponding to an identity may be seen as a signature by the PKG on the identity. In other words, the identity is considered to be the message and the decryption key is the signature of PKG on this message. Verification of the correctness of the decryption key (signature) can be done by encrypting a random string under the identity (message) and then using the decryption algorithm with the decryption key to see if the string is recovered properly.

Suppose now that the IBE scheme is secure against adaptive-identity attacks. In such an attack, the adversary submits identities and receives corresponding decryption keys. Viewed as a signature scheme, this would be submitting messages and receiving the corresponding signatures. Suppose now that the adversary can produce a valid forgery, i.e., the adversary can produce a valid message-signature pair. Switching back to the IBE scenario, the adversary can submit the message as the challenge identity and the corresponding signature is a decryption key for this identity. Knowing a valid decryption key for the challenge identity, the adversary can decrypt the challenge ciphertext and win the IBE security game. This simple observation of how an adaptive-identity secure IBE scheme provides a signature scheme has been attributed to Moni Naor in [39].

The above strategy has been applied to several IBE schemes. The first such scheme is by Boneh, Lynn and Shacham [45] and is called the BLS signature scheme. The importance of this scheme is two-fold. First, it is the first concrete instantiation of Naor's observation. Second, the scheme makes use of asymmetric pairings to obtain signatures that are quite small. Since signature schemes are extensively used in practice and a large number of signatures are stored and transmitted across networks, any reduction in the size of a signature is of great practical importance.

The BLS signature scheme makes use of a hash function $H : \{0,1\}^* \to G_1$ and an asymmetric pairing $e : G_1 \times G_2 \to G_T$. Let $G_1 = \langle P_1 \rangle$ and $G_2 = \langle P_2 \rangle$. The security analysis views H as a random oracle.

KeyGen. Pick random x from $\mathbb{Z}_p$ and compute $P_{pub} = xP_2$. The public key is the pair (P_2, P_{pub}). The private signing key is x.

Signing. Given a private key x and a message $M \in \{0,1\}^*$, compute $Q = H(M) \in G_1$ and $\sigma = xQ$. The signature is $\sigma \in G_1$.

Verify. Given a public key $(P_2, P_{pub}) \in G_1 \times G_2$, a message $M \in \{0,1\}^*$ and a signature $\sigma \in G_1$, compute $Q = H(M) \in G_1$ and verify that $(P_2, P_{pub}, Q, \sigma)$ is a valid co-Diffie-Hellman tuple, i.e., check whether the following condition holds:

$$e(\sigma, P_2) = e(Q, P_{pub}).$$

If so, output valid; otherwise, output invalid. The correctness of verification can be seen from the following computation.

$$e(\sigma, P_2) = e(xQ, P_2) = e(Q, xP_2) = e(Q, P_{pub}).$$

A signature is a single element of G_1. It is possible to choose parameters such that elements of G_1 have a short representation *and* a desired security is achieved. We refer to [45] for details of such choice.

Aggregation of signatures. Certain applications require managing signatures by different entities on separate messages. As an example, consider the scenario of certification authorities (CA) where one needs to manage certificate chains which contain signatures of CAs on distinct certificates. It would be convenient if there is some method which aggregates all the signatures into a single signature. A scheme which permits this is called an aggregate signature scheme. This notion was introduced by Boneh, Gentry, Lynn and Shacham [41] and they proposed the following technique for signature aggregation.

Suppose n users each have a public-private key pair of the Boneh-Lynn-Shacham signature scheme. So, for $i = 1, \ldots, n$, user i has private key $x_i \in \mathbb{Z}_p$ and public key $P_i = x_i P_2$. Suppose user i signs a message $M_i \in \{0,1\}^*$ to obtain the signature $\sigma_i = x_i H(M_i)$. These signatures are aggregated as follows: $\sigma = \sigma_1 + \sigma_2 + \cdots + \sigma_n$. Note that σ is an element of G_1. Verification of the aggregate signature is done in the following manner. Given the public keys $P_1, \ldots, P_n$ and the messages $M_1, \ldots, M_n$, and the aggregate signature σ, output valid if the following two conditions hold, otherwise output invalid.

1. The messages $M_1, \ldots, M_n$ are distinct.
2. $e(\sigma, P_2) = \sum_{i=1}^{n} e(H(M_i), P_i)$.

This verifies that the for each i, signatures σ_i is a valid signature of entity i on message M_i. The security analysis shows how an adversary against this scheme can be used to construct an adversary against the Boneh-Lynn-Shacham signature scheme. For details of the security analysis, we refer to [41].

10.1.2 A Hierarchical Identity-Based Signature

As in the case of PKE schemes, for a signature scheme to be properly functional, the verifier Bob needs to trust the verification key of the signer Alice, i.e., Bob needs to be sure that the public key indeed belongs to Alice. As otherwise, an adversary Eve may use her signing key to sign a message and then pose as Alice to Bob, i.e., present her verification key as that of Alice. This will lead to Bob accepting that the message has been signed by Alice when it has actually been signed by Eve. So, Bob needs some way to verify that Alice's verification key indeed belongs to her. This brings in the issue of digital certificates and certifying authorities as in the case of PKE schemes.

Identity-based signature (IBS) can avoid using the associated machinery required for a usual signature scheme to work. This notion was proposed by Shamir himself in his pioneering work on identity based encryption [155]. As in the case of IBE schemes, there is a PKG who publishes public parameters (PP) and possesses a master secret key. It issues signing keys associated to an identity. A signer uses his/her identity, the signing key and the PP to sign a message. The verifier uses the identity of the signer and the PP to verify a message-signature pair. Similar to encryption, this has a straightforward extension to a hierarchical identity-based signature scheme (HIBS).

It should be noted that constructing an IBS scheme is significantly easier compared to that of an IBE scheme. Shamir's original paper on identity-based cryptosystems [155] had itself provided an IBS scheme even though it took significantly more time to come up with an IBE scheme.

An HIBS scheme can be seen as an extension of a signature scheme to the identity-based setting. We provide the definition of such a scheme.

Definition. A HIBS scheme consists of four algorithms (which are probabilistic and polynomial time in the security parameter): Set-Up, KeyGen, Sign and Verify. For a HIBS of height h (henceforth denoted as h-HIBS) any identity id is a tuple $(\mathrm{id}_1, \ldots, \mathrm{id}_j)$ where $1 \leq j \leq h$.

- HIBS.SetUp and HIBS.KeyGen($\mathrm{id}, d_{\mathrm{id}|_{j-1}}, PP$) are exactly the same as that of an HIBE scheme.
- HIBS.Sign($\mathrm{id}, d_{\mathrm{id}}, M, PP$). Takes as input id, a private key d_{id} for id, the message M and the public parameter of the PKG PP, and returns sig, the signature of M under the identity id.
- HIBS.Verify($\mathrm{id}, M, \mathrm{sig}, PP$). Takes as input id, message M, signature sig and outputs yes if sig is a valid signature for M under id or if this does not hold, then it outputs no.

Note that for $h = 1$, the above definition reduces to that of an IBS scheme. The security model for existential unforgeability under chosen message attacks consists of a game between an adversary and a simulator that goes through the following phases.

Set-Up. The simulator sets up the HIBS scheme, i.e., generates the public parameter PP and the master secret key for the scheme and provides the adversary with PP.

Queries. The adversary makes two types of queries in an interleaved and adaptive manner.

- *Extract queries.* The adversary can ask for the private key of any identity. The simulator provides a private key for this identity and the distribution of the private key should be the same as that generated by HIBS.KeyGen.
- *Signature queries.* In this type of query, the adversary provides an identity and a message. The simulator has to provide a proper signature on the message under the given identity.

Forgery. At the end of the interaction, the adversary outputs a message M^*, an identity id^* and a signature sig^*. The adversary is successful if the followings hold.

- HIBS.Verify($\mathrm{id}^*, M^*, \mathrm{sig}^*, PP$) returns yes.
- The adversary has not made any previous key extraction query on id^* or any of its prefix.
- The adversary has not made any previous signature query on (M^*, id^*).

The advantage of an adversary is defined to be the probability that the adversary succeeds in the above game. As usual, the HIBS scheme is said to be $(t, q_{id}, q_S, \varepsilon)$-secure if the advantage of any adversary which runs in time t, makes q_{id} key extraction queries and q_S signature queries is at most ε.

In the above definition, one requirement on a forgery is that the pair (M^*, id^*) has to be "new". This condition can be changed by only insisting that the signature sig* is "new". For this changed condition, an adversary can be successful by producing a new signature on an earlier message. A signature scheme satisfying this new requirement is said to satisfy *strong* unforgeability against chosen message attacks (as opposed to *existential* unforgeability against chosen message attacks). But, the signature scheme obtained below does not satisfy this condition as it is possible to convert an already known signature on a message into a new signature on the same message.

Construction. The method of converting an IBE to a signature scheme extends to convert an HIBE scheme into a hierarchical identity-based signature (HIBS) scheme. This has been noted in [94].

An entity with identity id $= (id_1, \ldots, id_j)$ has a corresponding decryption key d_{id}. Signature on a message M is then simply the decryption key for the identity tuple $(id_1, \ldots, id_j, M)$ which can be generated using d_{id}, which now acts as a signing key. So an $(h+1)$-level HIBE scheme gives rise to an h-level HIBS scheme. Though conceptually simple, there is a problematic issue. For such a scheme, the message space and the identity space should be disjoint. As otherwise, someone possessing the signature of M for the identity $(id_1, \ldots, id_j)$ can now use this signature as a signing key to produce a forgery for a message M' under the identity $(id_1, \ldots, id_j, M)$.

Paterson and Schuldt [144] had described an identity-based signature (IBS) scheme based on 2-level Waters' HIBE in [169]. The scheme has been proved to be secure. Now, consider the situation where the individual entities *and* the PKG *both* sign messages. The PKG signs messages using the master secret key and an individual entity signs messages using the signing key obtained from the PKG. For such a scenario, the scheme in [144] is no longer secure. Basically, the above situation applies. An adversary may obtain the PKG's signature on a "message" M and then use this signature as the signing key to sign messages under the "identity" M. This happens due to the fact that the scheme in [144] does not ensure that the message and identity spaces are disjoint.

As mentioned above an HIBS scheme can be obtained from the Gentry-Silverberg HIBE [94]. The security of this scheme holds in the random oracle model. The technical details are essentially an extension of the way in which the BLS signature scheme is obtained from the BF-IBE. Below we describe the construction of a HIBS scheme which is based upon [63]. A signature scheme and an IBS scheme can be obtained as particular cases. The IBS improves upon [144] by reducing the size of the public parameters by almost half. This improvement is a result of the reduction in public parameters of the HIBE in Section 7.4 of Chapter 7 over the HIBE in [169].

Let $H : \{0, 1\}^* \rightarrow \{0, 1\}^n$ be a collision resistant hash function. The output of H is assumed to consist of l blocks where each block is an n/l-bit string considered to be an element of the set $\{0,\ldots,2^{n/l}-1\}$. Messages are assumed to be arbitrary binary strings and identities are assumed to be of the type $(\text{str}_1,\ldots,\text{str}_j)$, $1 \leq j \leq h$ and str_k, $1 \leq k \leq j$ is an arbitrary binary string. These are hashed into n-bit strings in the following manner. If msg is a message, then compute $H(0||\text{msg})$, while if str_k is a component of an identity tuple then compute $H(1||\text{str}_k)$. This ensures that n-bit strings obtained from messages will not be equal to n-bit strings obtained from identity components (assuming that H is collision resistant). Following the notation of HIBE schemes, we will write $\text{id}_k = H(1||\text{str}_k)$.

With the above modifications, one can easily convert an $(h+1)$ level HIBE to an h level HIBS. For an identity at the j-th level, the signature will be the decryption key of a $(j+1)$ level identity, with the message to be signed constituting the last level "identity" in the hierarchy.

The HIBS scheme can be described using both symmetric as well as asymmetric pairings. Since we have described the Boneh-Lynn-Shacham scheme using asymmetric pairings, we do the same for the HIBS scheme.

Set-Up. As in the HIBE scheme in Section 7.4, identities consists of l blocks of n/l-bit strings, where n is the security parameter and l is an integer which divides n.

The scheme is built from an asymmetric pairing $(G_1 = \langle P_1 \rangle, G_2 = \langle P_2 \rangle, G_T, e)$. Suppose the maximum number of levels in the HIBS is h. The PKG chooses random $x_i, y_j \in \mathbb{Z}_p^*$, where $1 \leq i \leq h+1$ and $1 \leq j \leq l$ and computes $U_i' = x_i P_1$, $W_i' = x_i P_2$, $U_j = y_j P_1$, $W_j = y_j P_2$. The PKG also chooses a random $R_1 \in G_1$ and a random integer $x \in \mathbb{Z}_p$ and computes $Q_2 = x P_2$ and $e(R_1, Q_2)$. The public parameters are the following elements: P_2, $e(R_1, Q_2)$, $U_1', \ldots, U_{h+1}', U_1, \ldots, U_l, W_1', \ldots, W_{h+1}', W_1, \ldots, W_l)$, The master secret is $x R_1$. The hash function H is also specified as part of the set-up.

KeyGen. Let $(\text{str}_1, \ldots, \text{str}_j)$ be the identity for which a key has to be generated and let $\text{id}_k = H(1||\text{str}_k)$ for $1 \leq k \leq j$. A key corresponding to this identity is generated by essentially applying the key generation algorithm of the HIBE scheme in Section 7.4 to $(\text{id}_1, \ldots, \text{id}_j)$. For example, for a first level identity str_1, $\text{id}_1 = H(1||\text{str}_1)$ and the PKG computes the corresponding private key as $d_0 = x R_1 + r_1 V_{1,1}(\text{id}_1)$ and $d_1 = r_1 P_1$, where $V_{1,k}(\text{id}_k) = U_k' + \sum_{i=1}^{l} \text{id}_{k,i} U_i$ for $1 \leq k \leq h+1$ and $r_1 \in_R \mathbb{Z}_p^*$. Similarly, the entity with a signing key for $(\text{id}_1, \ldots, \text{id}_{j-1})$ (i.e. for $(\text{str}_1, \ldots, \text{str}_{j-1})$) can generate a signing key for $(\text{id}_1, \ldots, \text{id}_j)$ (i.e. for $(\text{str}_1, \ldots, \text{str}_j)$). Note that the signing key thus generated contains elements of G_1 only.

Sign. Suppose a message msg is to be signed under an identity $(\text{str}_1, \ldots, \text{str}_j)$. Let $\text{id}_i = H(1||\text{str}_i)$, for $1 \leq i \leq j$ and $\text{id}_{j+1} = H(0||\text{msg})$ and let $\text{id} = (\text{id}_1, \ldots, \text{id}_j)$. Suppose that d_{id} is a signing key for id (i.e., for $(\text{str}_1, \ldots, \text{str}_j)$).

Then a signature on msg under the identity $(\text{str}_1, \ldots, \text{str}_j)$ is obtained by applying the key generation algorithm described above in the following manner. Using the key d_{id} for id, a key for the "identity" $(\text{id}_1, \ldots, \text{id}_j, \text{id}_{j+1})$ is created and this key is returned as the signature sig.

Verify. The input is a tuple $(\text{msg}, (\text{str}_1, \ldots, \text{str}_j), \text{sig})$ where

$$\text{sig} = (d_0, d_1, \ldots, d_{j+1}) \in G_1^{j+2}.$$

Let $id_k = H(1\|str_k)$, $1 \le k \le j$ and $id_{j+1} = H(0\|msg)$ and let $V_{2,k} = V_{2,k}(id_k) = W'_k + \sum_{i=1}^{l} id_{k,i}W_i$ for $1 \le k \le h+1$. The input is accepted if the following equality holds:

$$e(d_0, P_2) = e(R_1, Q_2) \times \prod_{k=1}^{j+1} e(d_k, V_{2,k}).$$

Note.

1. The security of this signature scheme can be shown to be based on the co-CDH problem in G_1 and G_2 for asymmetric pairing settings. This problem is the analogue of the CDH problem in G for symmetric pairing settings. See [45] for the definition of the co-CDH problem.
2. If $h = 1$, then the above scheme yields an identity-based signature (IBS).
3. The scheme can also be seen as providing a usual signature scheme. For this, the PKG is no longer required since there are no identities. Neither is the hash function H required and the messages are taken to be bit strings of length n. Further for $l = n$, the symmetric pairing based variant of this signature scheme is exactly the signature scheme described by Waters [169].

10.2 Identity-Based Key Agreement

In Chapter 1 we discussed the simple one-round two party key agreement protocol proposed by Diffie and Hellman in their seminal paper [77] published in 1976. Later another elegant protocol due to Burmester and Desmedt [50] showed how to perform key agreement between any number of parties in two rounds (assuming the existence of a broadcast channel).

A natural question is whether there is any *one*-round protocol for key agreement between three (or more) parties. This question eluded an answer until the new millennium when Joux [112] proposed a simple and elegant protocol for three-party one-round key agreement. This is regarded as the first application of bilinear pairings for constructing cryptographic primitives. Let $e : G \times G \to G_T$ be a pairing function in the symmetric setting and $G = \langle P \rangle$. Suppose there are three users U_1, U_2 and U_3. In the protocol user U_i selects a random x_i; computes $X_i = x_i P$ and sends X_i to the other two users. Then each user can compute the common key $e(P,P)^{x_1 x_2 x_3}$ as shown below.

$$e(P,P)^{x_1 x_2 x_3} = e(P_2, P_3)^{x_3} \text{ (User 1)}$$
$$= e(P_1, P_3)^{x_2} \text{ (User 2)}$$
$$= e(P_2, P_3)^{x_1} \text{ (User 3)}.$$

In Chapter 1 we also introduced the ElGamal scheme for public key encryption and its close relation with the Diffie-Hellman key agreement scheme. So another natural question would be whether there is any connection between identity-based key agreement and identity-based encryption. Before answering this question we discuss the first identity-based key agreement protocol proposed by Sakai, Ohgishi and Kasahara [151]. This protocol was also introduced in the year 2000, but was published in Japanese. Hence it took a few years before the international research community became aware of this equally simple and elegant protocol.

The first identity-based key agreement protocol that was proposed by Sakai, Ohgishi and Kasahara is non-interactive, i.e., two users A and B having identities id_A and id_B agree upon a common key without any interaction among themselves. As in IBE, there is a PKG who provides the private key corresponding to each identity. The protocol proceeds as follows:

Set-Up. Let $e : G \times G \to G_T$ be a pairing function in the symmetric setting and $G = \langle P \rangle$. Two hash functions $H : \{0,1\}^* \to G$ and $H_2 : G_T \to \{0,1\}^n$ are required. PKG chooses a random s from $\mathbb{Z}_p$ to be the master secret key and sets $R = sP$ and the hash function H_1, H_2 to be the public parameters.

Key Extraction. The private key d_A of user A with identity id_A is defined to be $d_A = sQ_A$, where $Q_A = H(id_A)$.

Key Agreement. A and B having identities id_A and id_B can compute a common key as follows. A computes $K_A = H_2(e(d_A, Q_B)) = H_2(e(Q_A, Q_B)^s)$ and B computes $K_B = H_2(e(d_B, Q_A)) = H_2(e(Q_B, Q_A)^s)$. Note that in the symmetric setting we have $e(Q_A, Q_B) = e(Q_B, Q_A)$. Hence, $K_A = K_B$ is the shared key of A and B.

Coming back to our original question of relationship between identity-based key agreement and IBE, let's compare the above scheme with the BasicIdent protocol of Boneh and Franklin. The Set-up and KeyGen algorithms are very similar. The Encrypt and Decrypt algorithms can be obtained by slightly tweaking the Key Agreement algorithm. Recall that in BasicIdent the message is masked by the hash of $e(P_{pub}, Q_{id})^r = e(rsP, Q_{id})$ where $r \in \mathbb{Z}_p$ is an ephemeral secret chosen by the sender and rP is sent as part of the ciphertext. The recipient with public key Q_{id} and private key $d_{id} = sQ_{id}$ unmasks the message by computing the hash of $e(rP, d_{id}) = e(srP, Q_{id})$. Now think of this rP as the "identity" of the sender and then it becomes clear that the shared secret in the Sakai-Ohgishi-Kasahara key agreement scheme is transformed into the mask created in the BasicIdent.

The above technique is quite similar to transforming the Diffie-Hellman key agreement to obtain the ElGamal encryption scheme and was discussed in Chapter 1. The above relationship between the Sakai-Ohgishi-Kasahara non-interactive key agreement and BasicIdent was pointed out in a recent work [145]. In fact, [145] shows that any identity-based non-interactive key agreement protocol satisfying certain technical conditions can be generically converted to an identity-based encryption protocol.

Sakai-Ohgishi and Kasahara also proposed an interactive version of their identity-based key agreement scheme. This was soon followed by several other protocols. See [64] for a unified description of identity-based key agreement protocols employing bilinear pairings. Later work on identity-based key agreement can be found in [65, 85].

10.3 Broadcast Encryption

Broadly speaking, the idea of broadcast encryption is to perform a single encryption of a message that can be decrypted by a number of recipients. The main issue is that any particular message may only be intended for a subset of the possible set of recipients. These recipients are called privileged (with respect to this message) while others are called revoked. Two crucial parameters are of interest. The first is the size of the broadcast and the second is the size of the key material that individual recipients need to store. Broadcast encryption has different applications which include pay TV systems, DVD content protection and access control in encrypted file systems.

The formal notion of broadcast encryption was introduced by Fiat and Naor [80]. This was mainly in the setting of symmetric key encryption. Public key broadcast encryption schemes were considered later [78]. The security model for public key broadcast encryption is an extension of the security model for usual PKE schemes. Differences arise due to the fact that a usual PKE scheme has a single recipient whereas a broadcast scheme has multiple recipients. For example, in a decryption query, the adversary specifies a recipient. Somewhat more importantly, for generation of the challenge ciphertext, the adversary specifies a target set of recipients.

From the point of view of obtaining a security reduction, a crucial issue is the manner in which the adversary is allowed to choose the target set of recipients. In the adaptive model, the adversary can choose this set after making decryption queries, while in the non-adaptive model, the adversary has to specify the target set at the outset of the security game. Correspondingly, it is much more difficult to obtain a scheme which can be proved secure in the adaptive model.

The above description should make the connection to identity-based techniques clear. The basic similarity is that both IBE and broadcast encryption deal with a number of recipients where each recipient has a private key which is unknown to the other recipients. Further, the adaptive and non-adaptive security models are roughly analogous to the adaptive-identity and selective-identity attacks for IBE security model.

The connection becomes clearer, when we consider the selective-identity security model for HIBE schemes. The adversary has to specify the target identity at the beginning of the security game. For an HIBE scheme, an identity is a tuple where each prefix of the tuple represents an individual possessing a private key. So, one may consider such an identity as specifying a possible set of recipients. From this, it is quite natural to try to obtain a broadcast encryption scheme secure against non-adaptive attacks from an HIBE scheme which is secure against selective-identity attacks.

A HIBE scheme, on the other hand has an additional structure in the form of the requirement of being able to support key delegation over levels. In other words, an identity tuple possessing a private key should be able to generate keys for identities for which it is a proper prefix. Such a requirement is not present in a broadcast encryption scheme. Moreover, the security model for the non-adaptive setting does not provide the adversary with a key-extraction oracle. Instead the adversary is provided with the private keys of the revoked users immediately after the system is set-up. To a certain extent, this makes the task of private key generation in a broadcast encryption scheme somewhat simpler than that of an HIBE scheme.

We have talked about broadcast encryption schemes secure against non-adaptive adversaries. By a similar reasoning, an HIBE which is secure against adaptive-identity attacks provides an idea of constructing a broadcast encryption scheme which is secure against adaptive adversaries. It is only recently that a somewhat satisfactory solution for adaptive-identity secure HIBE has been found [170]. As a natural extension, this has led to the construction of a broadcast encryption scheme which is secure against adaptive identities [170].

In a broadcast encryption scheme a broadcast message is typically of the form (S, Hdr, C_M), where S is the set of privileged recipients, Hdr is an encapsulation of a key K which is used to encrypt a message M through a symmetric encryption scheme to obtain the ciphertext C_M. Each user $u_i \in S$ uses her private key to decapsulate the key K from the Hdr and then uses K to decrypt the broadcast message.

Here we describe only one broadcast encryption scheme. This is the first broadcast encryption scheme which exploits the above mentioned connection to HIBE schemes. As described in Chapter 5, Boneh, Boyen and Goh had obtained an interesting construction of a selective-identity secure HIBE scheme possessing a constant size ciphertext. Following the above intuition, this can be modified (albeit in a non-trivial way) to obtain a broadcast encryption scheme. Boneh, Gentry and Waters [42] proposed such a construction. The crucial aspect of this scheme is that the constant size ciphertext for the HIBE scheme translates into a constant size Hdr for the broadcast encryption. A drawback is that the size of the public parameters is linear in the number of recipients. Since this number can be large, this may significantly affect practical implementation. To alleviate this problem, the paper [42] provides a mechanism for obtaining a controllable trade-off between the size of the public parameters and the size of the ciphertext. Below we describe the basic scheme from [42].

Set-Up. The description is in terms of a symmetric pairing $e : G \times G \to G_T$ with $G = \langle P \rangle$. To set up the scheme, choose a uniform random $\alpha \in \mathbb{Z}_p$ and compute $P_i = \alpha^i P$ for $i = 1, \ldots, 2n$, where n is the maximum number of recipients the system can support. Further, choose a random $\gamma \in \mathbb{Z}_p$ and set $Q = \gamma P$. The public key is the tuple

$$(P, P_1, \ldots, P_n, P_{n+2}, \ldots, P_{2n}, Q).$$

Note that the element P_{n+1} is not part of the public key. The private key for user $i \in \{1, \ldots, n\}$ is set to be $d_i = \gamma P_i$ and so $d_i = \alpha^i Q$.

Encrypt. Given a set S of privileged users, the header is generated as follows. Choose a random t in $\mathbb{Z}_p$ and set the secret key to be $K = e(P_{n+1}, P)^t = e(P_n, P_1)^t$. The header Hdr is defined to be

$$\left(tP, t\left(Q + \sum_{j \in S} P_{n+1-j}\right) \right).$$

Decrypt. Suppose u_i is a user and belongs to a set S of users for which the broadcast has been created. Let d_i be the private key of user u_i. Let Hdr $= (C_0, C_1)$. Then, the secret key K is reconstructed in the following manner.

$$\frac{e(P_i, C_1)}{e\left(d_i + \sum_{\substack{j \in S \\ j \neq i}} P_{n+1-j+i}, C_0 \right)}.$$

The correctness of decryption can be seen from the following computation.

$$\frac{e(P_i, C_1)}{e\left(d_i + \sum_{\substack{j \in S \\ j \neq i}} P_{n+1-j+i}, C_0 \right)} = \frac{e(\alpha^i P, t(Q + \sum_{j \in S} P_{n+1-j}))}{e\left(\alpha^i Q + \sum_{\substack{j \in S \\ j \neq i}} P_{n+1-j+i}, tP \right)}$$

$$= e(\alpha^i P, t P_{n+1-i}) \times \frac{e\left(\alpha^i P, t\left(Q + \sum_{\substack{j \in S \\ j \neq i}} P_{n+1-j}\right) \right)}{e\left(\alpha^i \left(Q + \sum_{\substack{j \in S \\ j \neq i}} P_{n+1-j}\right), tP \right)}$$

$$= e(P, t P_{n+1}) \times \frac{e\left(\alpha^i P, t\left(Q + \sum_{\substack{j \in S \\ j \neq i}} P_{n+1-j}\right) \right)}{e\left(t(Q + \sum_{\substack{j \in S \\ j \neq i}} P_{n+1-j}), \alpha^i P \right)}$$

$$= e(P, P_{n+1})^t$$

$$= K.$$

In the above scheme, a private key is only one group element and the ciphertext consists of two group elements. Since $e(P_{n+1}, P)$ can be precomputed, encryption requires no pairings. The drawback is that the size of the public parameters is almost twice the number of possible recipients. The system is able to broadcast to any subset of users and is fully collusion resistant against a non-adaptive adversary who is not allowed to make any decryption query. By composing this protocol with the Boneh-Boyen IBE (somewhat akin to what is done in Section 7.5 of Chapter 7) and applying the generic transformation discussed in Chapter 6 it is possible to obtain a CCA-secure scheme.

10.4 Fuzzy Identity-Based Encryption

An interesting extension of IBE was introduced by Sahai and Waters [150]. Suppose
the identity of a user is generated from his/her biometric data. The user receives a
decryption key from the PKG corresponding to this identity. Two biometric mea-
surements are rarely ever the same. So, it is possible that a message is encrypted
to an identity obtained from a measurement different from the one that yielded the
identity for which the decryption key was obtained. For usual IBE schemes, the user
will not be able to decrypt the message. Some amount of error-correction capability
needs to be built into the system.

In somewhat more concrete terms, a decryption key is obtained for an identity id
and the encryption is to an identity id$'$, then a fuzzy IBE scheme allows decryption
if id and id$'$ are "close". The notion of closeness is in the sense of error correct-
ing codes and not surprisingly, the construction of fuzzy IBE scheme in [150] is
based on Reed-Solomon codes stated in terms of Shamir's secret sharing scheme.
We provide a brief description of one of the schemes from [150]. For more details,
including the appropriate security model and proofs we refer the reader to the original paper [150].

Set-Up. A symmetric pairing $e : G \times G \to G_T$ is used where $G = \langle P \rangle$ and is of
order p. Identities are taken to be subsets of $\{1, \ldots, n\}$ and hence correspond to n-bit
strings. Choose $t_1, \ldots, t_n$ and y randomly from $\mathbb{Z}_p$ and compute $T_1 = t_1 P, \ldots, T_n =
t_n P$ and $Y = e(P,P)^y$. The public parameters of the PKG consist of $(T_1, \ldots, T_n, Y)$
and the master secret key is $(t_1, \ldots, t_n, y)$.

Key Extraction. Suppose id $\subseteq \{1, \ldots, n\}$ is an identity. A polynomial q of degree
$(d-1)$ is chosen such that $q(0) = y$. The decryption key consists of the elements
$(D_i)_{i \in \mathsf{id}}$ where $D_i = \frac{q(i)}{t_i} P$.

Encrypt. Suppose that a message $M \in G_T$ is to be encrypted to an identity id$'$. A
random s is chosen and the ciphertext is defined to be

$$E = (\mathsf{id}', E', (E_i)_{i \in \mathsf{id}'})$$

where $E' = M \times Y^s$ and $E_i = sT_i$. The identity id$'$ is included as part of the ciphertext
as it will be required during decryption.

Decrypt. Suppose that the ciphertext E is encrypted with id$'$ and the user has a
decryption key for id with id $\cap$ id$' \geq d$. Let S be a d-element subset of id $\cap$ id$'$.
Then the ciphertext can be decrypted as follows. The notation $\Delta_{i,S}(x)$ denotes the
Lagrange coefficient defined as $\Delta_{i,S}(x) = \prod_{j \in S, j \neq i}(x - j)/(i - j)$.

$$\frac{E'}{\prod_{i \in S} e(D_i, E_i)^{\Delta_{i,S}(0)}} = \frac{M \times e(P,P)^{sy}}{\prod_{i \in S} e(\frac{q(i)}{t_i}P, st_i P)^{\Delta_{i,S}(0)}}$$

$$= \frac{M \times e(P,P)^{sy}}{\prod_{i \in S} e(P,P)^{sq(i)\Delta_{i,S}(0)}}$$

$$= \frac{M \times e(P,P)^{sy}}{e(P,P)^{s\Sigma_{i \in S} q(i)\Delta_{i,S}(0)}}$$

$$= \frac{M \times e(P,P)^{sy}}{e(P,P)^{sy}}$$

$$= M.$$

Note that a polynomial reconstruction takes place in the exponent. This is where error tolerance comes into play.

10.5 Public Key Encryption with Keyword Search

A question of some practical importance is how to perform a keyword search on an encrypted message. One may consider that a user Alice has the option to read emails on a number of devices such as her desktop, laptop and mobile phone. Her mails come through a gateway which checks for the presence of a keyword. For example, if there is a keyword such as "urgent" then it is sent to Alice's mobile phone. The only problem is that if the message is encrypted, then how does the gateway determine this?

It is reasonable to assume that the number of keywords is small. The goal is to specify the keywords such that the gateway can search an encrypted mail for these keywords but, learn nothing else about the mail. This is achieved using a public key encryption with keyword search (PEKS). The broad view is the following. Suppose Bob wishes to send a message M with keywords $W_1, \ldots, W_m$. The transmission to Alice consists of the following.

$$E_{p_A}(M), \text{PEKS}(p_A, W_1), \ldots, \text{PEKS}(p_A, W_m)$$

where E denotes the encryption algorithm of a standard public key system and p_A is Alice's public key.

Alice gives the gateway a certain trapdoor T_W such that the gateway can test whether one of the keywords associated with the message is equal to the word W of Alice's choice. Applying T_W on $\text{PEKS}(p_A, W')$ the gateway can test whether $W = W'$. Apart from that the gateway learns nothing more about W'. For this search to take place Alice and Bob do not need to communicate. Alice provides the trapdoor to the gateway while Bob generates the searchable encryption for W using only the public key of Alice.

The notion of PEKS was introduced in [38] who gave a proper security model and a construction. A PEKS consists of four probabilistic algorithms.

1. KeyGen.
   Takes as input a security parameter s and generates a public/private key pair (p_A, s_A) for Alice.

2. PEKS(p_A, W).

   Given a public key p_A and a word W, the output S is a searchable encryption of W.

3. Trapdoor(s_A, W).

   Given Alice's private key and a word W produces a trapdoor T_W.

4. Test(p_A, S, T_W).

   Given Alice's public key, a searchable encryption $S = $ PEKS(p_A, W') and a trapdoor $T_W = $ Trapdoor(s_A, W), outputs 'yes' if $W = W'$ and 'no' otherwise.

The notion of PEKS is related to IBE scheme and in [38] it has been shown that a PEKS implies an IBE scheme where it is also conjectured that the converse is false. On the other hand, it is also known [2] that an anonymous IBE implies a PEKS. The argument that a PEKS implies an IBE is simple and can be stated as follows.

Given a PEKS an IBE system can be constructed as follows. Essentially, the public/private key pair (p_A, s_A) of any user Alice becomes the PP/master secret key of the PKG. An identity is an arbitrary binary string and is treated as a keyword W. The corresponding decryption key is a pair (d_0, d_1) of trapdoors for the keywords $W||0$ and $W||1$. Messages are taken to be bits. The ciphertext C corresponding to a message $b \in \{0, 1\}$ for an identity W, is the output of PEKS($p_A, W||b$). Decryption is done using the Test algorithm. If Test(p_A, C, d_0) returns 'yes', then return 0; and if Test(p_A, C, d_1) returns 'yes', then return 1. Note that exactly one of Test(p_A, C, d_0) or Test(p_A, C, d_1) will return 'yes'.

Even though a PEKS appears to be a more general primitive, the PEKS given in [38] is constructed essentially from the Boneh-Franklin IBE. We briefly describe the PEKS scheme from [38].

The setting is of symmetric pairing $e : G \times G \to G_T$ where $G = \langle P \rangle$ is of order p. Additionally, two hash functions $H_1 : \{0, 1\}^* \to G$ and $H_2 : G_T \to \{0, 1\}^{\log p}$ are required.

KeyGen. Choose a random $\alpha \in \mathbb{Z}_p^*$. Set $p_A = (P, Q = \alpha P)$ and $d_A = \alpha$.

PEKS(p_A, W). Choose a random $r \in \mathbb{Z}_p^*$ and compute $T = e(H_1(W), rQ)$. Output $(rP, H_2(T))$.

Trapdoor(p_A, W). Output $T_W = \alpha H_1(W)$.

Test(p_A, S, T_W). Let $S = (A, B)$. If $H_2(e(T_W, A)) = B$, then output 'yes', else output 'no'.

For proper T, S and T_W, the computation succeeds due to the following equality achieved through bilinear pairing.

$$e(T_W, A) = e(\alpha H_1(W), rP) = e(H_1(W), r\alpha P) = e(H_1(W), rQ) = T.$$

The scheme has been shown to be secure in the appropriate security model under the BDH assumption.

10.6 Other Applications

We briefly mention the connection of IBE based techniques to the construction of a few other primitives and further generalisations. Technical details are not provided. Only the motivation and a brief overview of the primitives are mentioned.

Wildcards. There are two ways to view wildcards depending on whether wildcards are used during encryption or during key delegation.

Identity-based encryption with wildcards (WIBE).

This notion was introduced in [3]. It extends the notion of HIBE. The key delegation feature of HIBE is unchanged. The difference is in the way encryption is formed. Suppose that a message is to be sent to all users whose identities are of the form $(\mathsf{id}_1, *, \mathsf{id}_3, \mathsf{id}_4)$ where $*$ matches any string. An example given in [3], is the hierarchy of email addresses at academic institutions. Suppose the requirement is to send encrypted emails to the following groups.

- *@cs.univ.edu, i.e., the mail is to be sent to all faculty members of the computer science department of some university univ.
- sysadmin@*.univ.edu, i.e., the mail is to be sent to the system administrators of all departments of univ. of the computer science department of some university univ.

Using a normal HIBE would require separate encryptions for each possible value of $*$. A WIBE provides the advantage of performing exactly one encryption to generate one ciphertext which is then sent to all the users. Any user possessing a decryption key for an identity of the above form (where the wildcard $(*)$ is replaced by a proper identity element) can then decrypt the ciphertext and recover the message.

Identity-based encryption with wildcard key derivation (WKD-IBE).

The complementary idea of using wildcards during encryption is to use wildcards during key delegation [4]. This primitive allows for more general forms of key delegation. In a WKD-IBE scheme, secret keys are generated for patterns rather than individual tuples. A pattern is a vector $(P_1, \ldots, P_n)$ where each P_i can be either an identity string or the special wildcard symbol $*$.

The general form of key delegation is the following. Let P and P' be patterns of lengths l and l' respectively. Suppose the following conditions hold.

1. $l' \leq l$;
2. for $i = 1, \ldots, l'$, either $P_i' = P_i$ or $P_i = *$;
3. $P_j = *$ for $j = l' + 1, \ldots, l$.

Then P' matches P. The aim of WKD-IBE is to ensure that an entity possessing a decryption key for P will be able to generate a decryption key for P'. For example, the head system administrator of a university may be given a key for the identity sysadmin@*.univ.edu. Then the head system administrator will be able to generate keys for the system administrator of any department. For this example, we would have $P = (\text{sysadmin}, *, \text{univ}, \text{edu})$ and P' can possibly be $(\text{sysadmin}, \text{cs}, \text{univ}, \text{edu})$ or $(\text{sysadmin}, \text{math}, \text{univ}, \text{edu})$. This can be easily seen to be an extension of the key delegation capability of HIBE.

Cryptographic Access control.
This is a broad and important problem in its own right. It refers to the ability of controlling access to sensitive data by different users. In an early paper, Smart [165] had shown how to use pairings to design an access control mechanism.

More sophisticated notions of access control have been developed. These come under the broad heading of attribute-based encryption. This notion was introduced by Sahai and Waters [150]. In such a scheme, a message is not encrypted to one particular user. One may consider a universal set of all possible attributes and a subset of this set can be considered to be a policy over attributes. Both decryption keys of the users as well as ciphertexts will be associated with a set of attributes. A user is able to decrypt a ciphertext if there is "significant" overlap between the attribute policy for the decryption key and the attribute policy for the ciphertext. One can clearly see the connection to the notion of fuzzy IBE scheme discussed above.

The above notion has been called key-policy attribute-based encryption. In such a scheme, the encryptor does not control who can decrypt the ciphertext except by the choice of the attributes used to encrypt the message. A related but different notion is that of ciphertext policy attribute-based encryption. Here, a user's decryption key is associated with an arbitrary number of attributes. The encryptor, specifies an access structure over the set of attributes. A user is able to decrypt a ciphertext only if that user's attributes pass through the ciphertext's access structure. The first construction of such a scheme was given in [29].

For a list of references on attribute-based encryption the reader can refer to [68].

Forward-secure encryption.
A constant threat in any cryptographic system is the possibility of compromise of the secret key. Usual encryption techniques offer no security if this occurs. Forward-secure encryption is an attempt to mitigate the effect of such compromise of static key. The idea is to ensure that compromise of long-term secret does not affect the secrecy of the previously encrypted messages. Anderson [11] proposed the problem of constructing a forward secure encryption scheme of the following type.

The lifetime of a system is divided into N intervals (or time periods), $0, \ldots, N-1$. The receiver starts with a key SK_0 which evolves over time periods. At the beginning of time period i, the receiver applies some function to the key SK_{i-1} of the previous time period to obtain the key SK_i of the current period. The key SK_{i-1} is then erased and SK_i is used. The public key, however, does not change during the lifetime.

From a security viewpoint, the requirement is that even if an adversary obtains SK_i, messages encrypted during all prior time periods remain secret. The system, on the other hand, clearly cannot protect the secrecy of messages encrypted at time period i and all subsequent time periods.

A formal security model for such a primitive was proposed in [53] who also presented a forward secure encryption scheme. The first step in this construction is a so-called binary tree encryption (BTE) which can be thought of as a primitive version of an HIBE. A BTE is then converted to a forward secure encryption scheme. From the above description, one may note the connection to an HIBE scheme. A vector $(0,2,\ldots,i)$, $i < N$ can be thought to be an identity tuple of an HIBE scheme and the key corresponding to this identity is defined to be SK_i. By the key delegation property of an HIBE scheme, possession of a secret key SK_{i-1} for the tuple $(1,2,\ldots,i-1)$ allows the generation of a secret key SK_i for the tuple $(0,\ldots,i)$. If the previous key SK_{i-1} is erased, then again by the property of a HIBE scheme, it will not be possible to generate SK_{i-1} from SK_i.

The notion of forward security has also been applied to (H)IBE schemes [172]. The goal is to mitigate possible damage caused by the exposer of the secret key in a (H)IBE scheme. Each user in the hierarchy will be able to refresh his/her decryption keys periodically without changing the public key.

10.7 Conclusion

This chapter provided an overview of the many different primitives that can be built using techniques derived from HIBE schemes. A signature scheme can be obtained generically while the relationship to other primitives such as public-key broadcast encryption is indirect. Several different extensions of the notion of IBE has also been discussed. As understanding of the algebraic techniques progresses there may result in the proposal of newer primitives and further applications of IBE techniques.

Chapter 11
Avoiding Key Escrow

A major issue with the deployment of IBE schemes is that the PKG is able to obtain a secret key of any user and hence will be able to decrypt any ciphertext. This is called the problem of inherent key escrow in IBE schemes. Methods to avoid this problem have been discussed in the literature. Below we briefly discuss some of these methods.

11.1 Distributed PKG

One way of avoiding key escrow is to use techniques of threshold cryptography. The master secret key is distributed across several PKGs and a minimum number of PKGs must cooperate to generate the secret key corresponding to an identity.

Let us consider this possibility for the BF-IBE. In this IBE, the master-key is some $s \in \mathbb{Z}_p$. The private key for an identity id is $d_{id} = sQ_{id}$, where Q_{id} is derived from id. This can easily be distributed in a t-out-of-n fashion by giving each of the n PKGs one share s_i of a Shamir secret sharing of $s \bmod p$.

Now, t PKGs can cooperate to generate a private key in the following manner. Each of the t PKGs provide $d_{id,i} = s_i Q_{id}$ to the user. The user then constructs $d_{id} = \sum \lambda_i d_{id,i}$ where the λ_i's are the appropriate Lagrange coefficients.

The fact that DDH is easy in G provides security against dishonest PKGs. During set-up, the i-th PKG ($1 \le i \le n$) publishes the public parameter $P_{pub,i} = s_i P$. This allows a user to test the validity of the key share provided by the i-th PKG by checking that the following computation holds: $e(d_{id,i}, P) = e(Q_{id}, P_{pub,i})$. This ensures that an erroneous key share will be immediately detected.

An additional feature is that the distributed master-key allows threshold decryption on a per-message basis, without directly deriving the corresponding decryption key. For example, threshold decryption of ciphertext (U, V) in the above scheme is straightforward if each PKG provides $e(s_i Q_{id}, U)$.

The idea of distributed PKG has been explored further in the literature and the CCA-security of Boneh-Franklin IBE, Sakai-Kasahara IBE and BB-IBE has been proved in the distributed PKG setting in [117].

11.2 Certificate-Less Encryption

A simple idea of avoiding key escrow in an IBE scheme is to use double encryption. A user submits its identity to the PKG and obtains a corresponding decryption key. Additionally, suppose that the user also uses a PKE scheme to choose a public key/private key pair. Encryption of a message to the user is done in two steps. In the first step, the message is encrypted using the public key of the user. Then, the ciphertext obtained as the output of the first step is encrypted using the encryption algorithm of the IBE scheme. This requires the PP of the PKG and the identity of the user. The resulting ciphertext is sent which the recipient can decrypt in two steps. First by decrypting using the decryption key of the IBE obtained from the PKG and then decrypting the output of the first step using the private key corresponding to its own public key. (It is also possible to envisage the reverse order of encryption, i.e., first using the IBE scheme and then using the PKE scheme.)

Intuitively, this two-fold strategy combines the best of both kinds of security of IBE and PKE schemes. Due to the encryption using the public key of the user, the PKG will no longer be able to decrypt the message. Using the encryption algorithm of the IBE scheme ensures the required binding of the identity to the PP of the PKG and alleviates concerns of trusting the identity of the user. There are two drawbacks to this simple approach.

1. The outer encryption operates on a string which is longer than the message length. This is because, the inner encryption produces an output which is longer than the plaintext.
2. The repeated encryption makes the encryption and decryption algorithms significantly slower compared to the individual PKE or IBE schemes. One goal of a concrete CL-PKE scheme is to avoid this slow down.

The basic idea of certificate-less public-key encryption is rooted in the above intuition. It is an attempt to obtain the advantages of avoiding the cumbersome public key infrastructure while getting rid of the issue of key escrow that is built into IBE schemes. Al-Riyami and Paterson [9] formally introduced this notion and provided constructions which avoid the efficiency loss of the simple idea outlined above.

A brief description of certificate-less encryption follows [9]. There is a trusted third party (TTP) whose role is different from that of a CA for PKE or of a PKG for an IBE. The TTP is called the key generating centre (KGC). Unlike the PKG, the KGC does not have access to the private keys of the users. The KGC supplies a user A with a partial private key d_A which the KGC computes from an identity id_A for the user and a master key. The task of providing partial private keys requires a secure channel between the user and the KGC. User A then combines its partial private key

d_A with some secret information to generate its actual private key s_A. Consequently, A's private key is not available to the KGC. User A combines its secret information with the KGC's public parameters to compute its public key P_A. The public key P_A for A may be made available by placing it in a public directory. Beyond this there is no additional security requirement to protect P_A and more specifically, there is no certificate for A's public key. To send an encrypted message to A, both P_A and id_A are required.

A CL-PKE scheme is specified by seven probabilistic algorithms: Set-Up, Partial-Private-Key-Extract, Set-Secret-Value, Set-Private-Key, Set-Public-Key, Encrypt and Decrypt. See [76, 66] for a survey.

The security model for a CL-PKE is an extension of the security model for usual IBE schemes. Complexities arise due to the more flexible kinds of attacks that can be mounted. Intuitively, in a CL-PKE, the public key of a user is no longer certified by a certifying authority. Thus, it is possible for an adversary to mount a typical masquerading attack, i.e., an adversary Eve may present her public key as that belonging to a user Alice. The security game models this by allowing the adversary to replace the public key of any user. The second kind of attack is that mounted by the KGC of the IBE scheme. Note that a goal of CL-PKE is to avoid key escrow. So, the security model has to consider the possibility of a KGC behaving in a malicious manner and trying to decrypt a ciphertext.

A natural restriction, however, is to disallow an adversary, which models a malicious KGC, from replacing public keys. For, if this is allowed, then the KGC can replace the public key of a user with a public key for which it knows a corresponding private key. As a consequence, any ciphertext intended for this user can be decrypted by the KGC – it can obtain a decryption key corresponding to the user's identity and also generate a private key for the fake public key. So, if both the attacks are simultaneously allowed, the adversary will always win the security game. The restriction mentioned above avoids this trivial situation.

Viewed from a different angle, the above situation can also be seen as an inherent weakness of CL-PKE schemes. One may argue that if an adversary is allowed to replace public keys, then, there is no reason not to allow the KGC to replace public keys. So, CL-PKE does not necessarily avoid the key escrow issue, possibly it only makes it somewhat more complicated for a malicious KGC to decrypt ciphertexts of a target identity.

Another related issue is that of denial-of-decryption attacks. Suppose that the KGC is not malicious. But, an adversary can still replace public keys of users. Now, if an adversary replaces the public key of a user, then an encrypted mail sent to this user cannot be decrypted by the user. Since the encryption also involves the identity of the user, neither can the adversary decrypt the ciphertext. But, if the adversary's goal is simply to deny the user from getting the message, then it would have succeeded.

The above discussion suggests that CL-PKE may not be the complete answer to the key escrow issue in IBE. Nevertheless, it represents an important step in addressing the problem. A modification of the BF-IBE to yield a CL-PKE was given by Al-Riyami and Paterson in [9] and is given below. Here the BF-IBE refers to the BasicIdent scheme in [39].

Set-Up. The setting is of symmetric pairing, where $G = \langle P \rangle$. Choose the master secret key uniformly at random from $\mathbb{Z}_p^*$ and set $P_{pub} = sP$. Let $H_1 : \{0,1\}^* \to G^*$ and $H_2 : G_T \to \{0,1\}^n$ be hash functions which the security analysis considers as random oracles. Here n refers to the length of the message that is to be encrypted.

Partial-Private-Key-Extract. This algorithm is the same as the key generation algorithm of the BF-IBE. Given an identity id, this algorithm returns $sH_1(\text{id})$.

Set-Secret-Value. This algorithm is run by a user and it simply returns a uniform random x from $\mathbb{Z}_p^*$.

Set-Private-Key. This algorithm combines the partial private key returned by the KGC and the secret value particular to a user to produce the actual secret key of the user. If d_{id} is the partial private key of a user with identity id having secret value x, then the actual secret key is returned to be $s_{\text{id}} = x d_{\text{id}} = xsH_1(\text{id})$.

Set-Public-Key. This algorithm sets the public key of a user using the public parameter of the KGC and the secret value of the user. If x is the secret value of a user, then the public key of the user is set to be $(xP, xP_{pub}) = (X, Y)$.

Given P and P_{pub} it is possible to determine the correctness of (X, Y) by the checking whether $e(P, Y) = e(P_{pub}, X)$. Such checking, however, only needs to be done once for each user.

Encrypt. The encryption of an n-bit message M with identity id and public key (X, Y) is done as follows: check that the equality $e(X, P_{pub}) = e(Y, P)$ holds, if not output $\perp$ and halt; choose a uniform random $t \in \mathbb{Z}_p^*$; set the ciphertext to be

$$(tP, M \oplus H_2(e(H_1(\text{id}), Y)^t)).$$

Decrypt. Given a ciphertext (U, V), decryption using a secret key s_{id} returns $V \oplus H_2(e(s_{\text{id}}, U))$.

In a manner similar to that of BF-IBE, it is easy to verify the correctness of decryption. The scheme is almost as efficient (in time and space) as the BF-IBE. A more general scheme which is analogous to the FullIdent scheme in [39] is shown to be CCA-secure in [9] using the hardness of the generalised BDH problem.

11.3 Certificate-Based Encryption

As already discussed, inherent key escrow is a basic problem with the wide-spread deployment of IBE. The other problematic issue is the secure transmission of the decryption keys to the users from the PKG. This requires a secure channel between the PKG and the user. Gentry [90] introduced the notion of certificate-based encryption (CBE) to solve both the problems at one go.

As in PKE, in a CBE, each user generates his/her own public-key/secret-key pair and requests a certificate from the CA. The crucial difference is that the CA uses an IBE scheme to generate the certificate.

Roughly speaking, the CA treats the user's public key as the identity and generates the secret key corresponding to it. Recall that, using Naor's observation, this secret key can be considered to be the CA's signature on the public key of the user. So, the public key-signature pair can be considered to be a certificate issued by the CA to the user. As a consequence, this certificate has the functionalities of a conventional PKI certification. Additionally, it can also be used as a decryption key. It is this added functionality which allows implicit certification. Now Alice can doubly encrypt a message to Bob so that Bob needs both his personal secret key and the certificate to decrypt.

There is no key escrow, since the CA does not know Bob's secret key; and there is no requirement for having a secure channel between the CA and Bob, since the CA's certificate need not be secret.

A concrete CBE scheme based on the BF-IBE was described in [90]. As in the case of BF-IBE, this comes in two variations – BasicCBE which is CPA-secure and FullCBE which is CCA-secure. Below we describe the BasicCBE scheme.

Set-Up. The setting is that of symmetric pairing with $G = \langle P \rangle$ and two hash functions $H_1 : \{0,1\}^* \to G$ and $H_2 : G_T \to \{0,1\}^n$, where n is the length of the message to be encrypted.

The set-up for the CA consists of choosing a uniform random secret key $s_C \in \mathbb{Z}_p$ and declaring public parameters to be $P_{pub} = s_C P$.

The set-up for an individual user Bob consists of choosing a secret key s_B and setting the public key for Bob to be $s_B P$.

Certification. Bob obtains certification from the CA in the following manner. Bob sends infoBob to the CA with his public key $s_B P$ and other personal identifying information. The CA computes $Q_B = H_1(P_{pub}, \text{infoBob})$ and returns the certificate $\text{cert}_B = s_C Q_B$ to Bob. If the CA wishes, then it may include a time-stamp period in the argument to H_1.

Bob now signs infoBob to produce $s_B P_B$ where $P_B = H_1(\text{infoBob})$ and constructs his secret key to be $S_B = \text{cert}_B + s_B P_B$. Note that this is essentially an *aggregation* of the CA's signature and Bob's signature on the two messages Q_B and P_B respectively.

Encrypt. Encryption of a message M using infoBob is done as follows: compute $Q_B = H_1(P_{pub}, \text{infoBob})$ and $P_B = H_1(\text{infoBob})$; choose a uniform random $t \in \mathbb{Z}_p$ and set the ciphertext to be $(tP, M \oplus H_2(e(P_{pub}, Q_B)e(s_B P, P_B)^t))$.

Decrypt. Given a ciphertext (U, V) encrypted using infoBob, decryption using the secret key S_B returns $V \oplus H_2(e(U, S_B))$.

Correctness of decryption follows from the following computation.

$$e(U, S_B) = e(tP, \text{cert}_B + s_B P_B)$$
$$= e(tP, \text{cert}_B) \times e(tP, s_B P_B)$$
$$= e(tP, s_C Q_B) \times e(tP, s_B P_B)$$

$$= e(s_C P, Q_B)^t \times e(s_B P, P_B)^t$$
$$= (e(P_{pub}, Q_B) \times e(s_B P, P_B))^t.$$

Note that the mask $(e(P_{pub}, Q_B)e(s_B P, P_B))^t$ can be seen as a combined randomisation by the two quantities $e(P_{pub}, Q_B)^t$ and $e(s_B P, P_B))^t$. This implicitly corresponds to a double encryption of the message M. Also, one may note how the aggregated signature which serves as the private key for Bob is used during decryption.

Gentry [90] describes several extensions of this basic scheme and provides an appropriate security model for CBE with security argument of the proposed schemes in that model. Later work [10] has examined the security model for CBE and has argued that a CBE scheme can be generically obtained from a certificate-less encryption scheme.

An important practical issue is that of key revocation. In the case of IBE and certificate-less encryption, handling key revocation is problematic. For CBE schemes, this becomes somewhat easier. The CA issues short-lived certificates at the beginning of every time period. This increases the computational overhead of the CA. Gentry [90] proposes some refinements of CBE which greatly reduces the CA workload. Another approach to tackling the key revocation problem is to use so-called mediated decryption. See [69] for a discussion on this aspect.

11.4 Other Approaches

Handling key escrow using accountable authority.
The problem of inherent key escrow has been viewed from a different and weaker perspective. In [100], the threat model considered is that of a malicious PKG selling off a decryption key for an identity. It is suggested that somebody may bribe the PKG to supply a decryption for some particular identity. Suppose further, that at a later stage, this decryption key is discovered by the genuine user. Now the question is whether the user can hold the PKG accountable for behaving maliciously.

As is clear, this question does not actually address the issue of key escrow. Using the master secret, the PKG can keep on decrypting each and every message that has been encrypted using its public parameters. The only disincentive to the PKG is that if it generates another decryption key which is discovered later, then it will be held accountable. This seems to guard against a rather weak threat model.

The idea introduced by Goyal in [100] is to have an exponential number of decryption keys corresponding to an identity. Key extraction process is modified to also involve the user and at the end of the process, the user gets a decryption key but, the PKG does not get to know which key the user has obtained. So, if the PKG later generates another decryption key for the same identity, then with high probability this key will be different from the one obtained by the genuine user. So, the simultaneous occurrence of two distinct decryption keys for a single identity is suggested to be taken as malicious behaviour on part of the PKG. For more details of how this is achieved, the reader may refer to [100] and the follow-up works in [101, 110].

Handling key escrow using anonymity.

Another approach to tackling key escrow has been suggested by Chow in [67]. In this approach, the idea is to use an IBE scheme where given a ciphertext, the PKG is unable to determine the identity under which it has been generated and hence the PKG does not know which private key is to be generated for decrypting this ciphertext. Put differently, the IBE scheme satisfies the notion of PKG-anonymous ciphertext indistinguishability. The formal model for this is introduced in [67] and it is shown that Gentry's IBE scheme can be modified to achieve this notion.

This notion, though interesting, has its limitations. For one thing, the method will be defeated if the identity is transmitted along with the ciphertext. Secondly, the PKG can always generate decryption keys for every user and attempt decryption of the ciphertext with each possible decryption key. The assumption in [67] is that there is a minimum amount of uncertainty in the PKG's knowledge about the set of intended recipients and it would be infeasible for the PKG to try out all possibilities. We refer the reader to the paper [67] for further details.

11.5 Conclusion

The issue of avoiding the inherent key escrow in IBE schemes has been discussed. For wide-scale deployment of IBE schemes, it is absolutely essential that this issue is properly addressed. Among the several approaches that have been proposed, only the certificate-less approach has been investigated in some detail. The certificate-based approach may be promising, but, requires further study. New approaches for tackling key escrow, especially in the context of HIBE schemes may also be developed.

Chapter 12
Products and Standards

It is slightly more than a decade ago that the first cryptographic applications of bilinear pairings were discovered. Extensive research has been done over the last decade on various aspects of such maps and the cryptographic primitives built from them. To a limited extent, this research has been implemented into concrete products for general use. This has been supplemented by the incorporation of pairing based schemes into proposed standards. The purpose of the current chapter is to provide a brief summary of the products and the draft standards related to identity-based cryptography.

To the best of our knowledge, there have been two commercial implementations of identity based cryptographic techniques. One is by Voltage Security Inc. of USA and the other is by an European company Identum which has been later acquired by TrendMicro. Among standards, there is a draft standard of the IEEE P1363.3/D1 for "Identity-based Public-key Cryptography Using Pairings" and a document by the Internet Engineering Task Force (IETF). These are discussed below.

12.1 Voltage Security

A commercial product for encrypted e-mails has been developed by Voltage Security Inc. This is based on the Boneh-Franklin IBE scheme and Professor Dan Boneh is one of the founders of the company.

As usual, identities can be arbitrary binary strings. To prevent against key compromise, a combination of a name and the validity period is used as the identity. An example of a key generated for Bob and valid from 1/1/2011 to 31/12/2011 would be of the form: "name=Bob validity=1/1/11-12/31/11". Since, the public key is different for each period, so is the corresponding decryption key. This mitigates the effect of compromise of a key to a certain period. Further, the management of identities can be quite flexible. For example, a policy may be enforced by including the policy statement as part of the identity.

12.2 Trend-Micro

An application for secure e-mail encryption based on the Sakai-Kasahara identity based key encapsulation mechanism has been commercialised. This was initially done by a company called Identum which was later taken over by a company called Trend-Micro. It appears that the actual scheme that has been implemented is a variant of the Sakai-Kasahara scheme.

12.3 IEEE P1363.3/D1 Draft Standard

The description below is from a draft standard and the usual caveat that the final standard may be different applies. Three IBE schemes are in the draft standard: Sakai-Kashara, Boneh-Boyen-1 and Boneh-Franklin.

The draft standard specifies identity-based cryptographic scheme based on bilinear maps. It is mentioned that the purpose of the draft is not to mandate any particular set of identity-based techniques, but, rather to provide a reference for specifications of a variety of techniques from which applications may select. The motivation stems from the fact that identity-based techniques can provide more efficient and convenient protocols. The draft standard defines the following types of schemes.

1. Identity-based Encryption.
2. Identity-based Key Encapsulation.
3. Identity-based Signature.
4. Identity-based Signcryption.

The use of these schemes is envisaged to allow communication between several parties in the following manner. Existence of one (or more) trusted key servers is required. These would correspond to what is more customarily called private key generator (PKG) in research papers. The schemes allow the sending party to convert an identity of the receiving party into a public key. This public key can be used to encrypt data. The receiving party has to request a decryption key for its identity from a key server. This decryption key is obtained only once and can be stored for subsequent use. To ensure security, a user needs to be sure of the ownership of the keys and domain parameters and of their validity. Proper generation of domain parameters and keys and check on their validity is essential. Appropriate key management is required but, it is mentioned that this is outside the scope of the draft standard.

A summary of the schemes included in the standard is given below.

1. Sakai-Kashara key encapsulation mechanism.
2. A key encapsulation mechanism based on the first Boneh-Boyen IBE scheme discussed in Chapter 5.
3. The Boneh-Boyen-IBE scheme itself is also included.

4. The Boneh-Franklin IBE scheme.
5. There are several other schemes including signature, signcryption, key agreement and proxy re-encryption methods.

Note that, the hash functions used in all the above IBE schemes (including the two variants of Boneh-Boyen) are modelled as random oracles.

For more details, we refer the reader to the actual draft [108].

12.4 IETF Memo

An IETF memo (RFC 5091) authored by Boyen and Martin of Voltage Security is available from [48]. It is mentioned that the memo does not specify an Internet standard of any kind and provides information for the Internet community. The document specifies two algorithms for identity-based encryption and is partly based on Voltage Security's Identity-based Cryptography Standards (IBCS). The algorithms are the Boneh-Franklin IBE scheme and the first Boneh-Boyen IBE scheme.

12.5 Conclusion

We have provided a very brief idea of the work that has been done on commercialising IBE schemes and developing standards. Hopefully, the coming years will see different kinds of IBE based applications arriving in the market. The ultimate success or failure of the notion of IBE as an important security technology will be judged by the extent of its deployment in solving real-life problems.

Chapter 13
Bibliography

References

1. M. Nakabayashi A. Miyaji and S. Takano. New explicit conditions of elliptic curve traces for FR-reduction. *IEICE Transactions on Fundamentals E84-A*, 5:1234–1243, 2001.
2. Michel Abdalla, Mihir Bellare, Dario Catalano, Eike Kiltz, Tadayoshi Kohno, Tanja Lange, John Malone-Lee, Gregory Neven, Pascal Paillier, and Haixia Shi. Searchable encryption revisited: Consistency properties, relation to anonymous IBE, and extensions. In Shoup [161], pages 205–222.
3. Michel Abdalla, Dario Catalano, Alexander W. Dent, John Malone-Lee, Gregory Neven, and Nigel P. Smart. Identity-based encryption gone wild. In Michele Bugliesi, Bart Preneel, Vladimiro Sassone, and Ingo Wegener, editors, *ICALP (2)*, volume 4052 of *Lecture Notes in Computer Science*, pages 300–311. Springer, 2006.
4. Michel Abdalla, Eike Kiltz, and Gregory Neven. Generalized key delegation for hierarchical identity-based encryption. In Joachim Biskup and Javier Lopez, editors, *ESORICS*, volume 4734 of *Lecture Notes in Computer Science*, pages 139–154. Springer, 2007.
5. Shweta Agrawal, Dan Boneh, and Xavier Boyen. Efficient lattice (H)IBE in the standard model. In Gilbert [95], pages 553–572.
6. Shweta Agrawal, Dan Boneh, and Xavier Boyen. Lattice basis delegation in fixed dimension and shorter-ciphertext hierarchical IBE. In Tal Rabin, editor, *CRYPTO*, volume 6223 of *Lecture Notes in Computer Science*, pages 98–115. Springer, 2010.
7. Miklós Ajtai. Generating hard instances of lattice problems (extended abstract). In *STOC*, pages 99–108, 1996.
8. Miklós Ajtai. Generating hard instances of the short basis problem. In Jirí Wiedermann, Peter van Emde Boas, and Mogens Nielsen, editors, *ICALP*, volume 1644 of *Lecture Notes in Computer Science*, pages 1–9. Springer, 1999.
9. Sattam S. Al-Riyami and Kenneth G. Paterson. Certificateless public key cryptography. In Chi-Sung Laih, editor, *ASIACRYPT*, volume 2894 of *Lecture Notes in Computer Science*, pages 452–473. Springer, 2003.
10. Sattam S. Al-Riyami and Kenneth G. Paterson. CBE from CL-PKE: A generic construction and efficient schemes. In Serge Vaudenay, editor, *Public Key Cryptography*, volume 3386 of *Lecture Notes in Computer Science*, pages 398–415. Springer, 2005.
11. Ross J. Anderson. Two remarks on public-key cryptology, invited lecture. In *4th ACM Conference on Computer and Communications Security*, 1997.
12. Diego F. Aranha, Julio López, and Darrel Hankerson. High-speed parallel software implementation of the Tate pairing. In Josef Pieprzyk, editor, *CT-RSA*, volume 5985 of *Lecture Notes in Computer Science*, pages 89–105. Springer, 2010.

13. Nuttapong Attrapadung, Yang Cui, David Galindo, Goichiro Hanaoka, Ichiro Hasuo, Hideki Imai, Kanta Matsuura, Peng Yang 0002, and Rui Zhang 0002. Relations among notions of security for identity based encryption schemes. In José R. Correa, Alejandro Hevia, and Marcos A. Kiwi, editors, *LATIN*, volume 3887 of *Lecture Notes in Computer Science*, pages 130–141. Springer, 2006.

14. Nuttapong Attrapadung, Jun Furukawa, Takeshi Gomi, Goichiro Hanaoka, Hideki Imai, and Rui Zhang 0002. Efficient identity-based encryption with tight security reduction. *IEICE Transactions*, 90-A(9):1803–1813, 2007.

15. Daniel V. Bailey and Christof Paar. Optimal extension fields for fast arithmetic in public-key algorithms. In Krawczyk [125], pages 472–485.

16. R. Balasubramanian and Neal Koblitz. The improbability that an elliptic curve has subexponential discrete log problem under the Menezes-Okamoto-Vanstone algorithm. *J. Cryptology*, 11(2):141–145, 1998.

17. Paulo S. L. M. Barreto. Pairing-based crypto lounge – a compendium of bilinear pairing related works. Available at: http://paginas.terra.com.br/informatica/paulobarreto/pblounge.html.

18. Paulo S. L. M. Barreto, Ben Lynn, and Michael Scott. On the selection of pairing-friendly groups. In Mitsuru Matsui and Robert J. Zuccherato, editors, *Selected Areas in Cryptography*, volume 3006 of *Lecture Notes in Computer Science*, pages 17–25. Springer, 2003.

19. Paulo S. L. M. Barreto and Michael Naehrig. Pairing-friendly elliptic curves of prime order. In Bart Preneel and Stafford E. Tavares, editors, *Selected Areas in Cryptography*, volume 3897 of *Lecture Notes in Computer Science*, pages 319–331. Springer, 2005.

20. Lynn Margaret Batten and Reihaneh Safavi-Naini, editors. *Information Security and Privacy, 11th Australasian Conference, ACISP 2006, Melbourne, Australia, July 3-5, 2006, Proceedings*, volume 4058 of *Lecture Notes in Computer Science*. Springer, 2006.

21. Mihir Bellare, Anand Desai, David Pointcheval, and Phillip Rogaway. Relations among notions of security for public-key encryption schemes. In Krawczyk [125], pages 26–45.

22. Mihir Bellare and Chanathip Namprempre. Authenticated encryption: Relations among notions and analysis of the generic composition paradigm. In Tatsuaki Okamoto, editor, *ASI-ACRYPT*, volume 1976 of *Lecture Notes in Computer Science*, pages 531–545. Springer, 2000.

23. Mihir Bellare and Thomas Ristenpart. Simulation without the artificial abort: Simplified proof and improved concrete security for waters' IBE scheme. In Antoine Joux, editor, *EUROCRYPT*, volume 5479 of *Lecture Notes in Computer Science*, pages 407–424. Springer, 2009.

24. Mihir Bellare and Phillip Rogaway. Random oracles are practical: A paradigm for designing efficient protocols. In *ACM Conference on Computer and Communications Security*, pages 62–73, 1993.

25. Mihir Bellare and Phillip Rogaway. Collision-resistant hashing: Towards making UOWHFs practical. In Kaliski-Jr. [116], pages 470–484.

26. Mihir Bellare and Phillip Rogaway. The security of triple encryption and a framework for code-based game-playing proofs. In Vaudenay [166], pages 409–426.

27. Daniel J. Bernstein. Pippenger's exponentiation algorithm. cr.yp.to/papers/pippenger.ps.

28. Daniel J. Bernstein and Tanja Lange. Explicit formulas database. http://www.hyperelliptic.org/EFD/.

29. John Bethencourt, Amit Sahai, and Brent Waters. Ciphertext-policy attribute-based encryption. In *IEEE Symposium on Security and Privacy*, pages 321–334. IEEE Computer Society, 2007.

30. Jean-Luc Beuchat, Jorge Enrique Gonzlez Daz, Shigeo Mitsunari, Eiji Okamoto, Francisco Rodrguez-Henrquez, and Tadanori Teruya. High-speed software implementation of the optimal Ate pairing over Barreto-Naehrig curves. Cryptology ePrint Archive, Report 2010/354, 2010. http://eprint.iacr.org/.

31. Eli Biham, editor. *Advances in Cryptology - EUROCRYPT 2003, International Conference on the Theory and Applications of Cryptographic Techniques, Warsaw, Poland, May 4-8, 2003, Proceedings*, volume 2656 of *Lecture Notes in Computer Science*. Springer, 2003.

32. Dan Boneh and Xavier Boyen. Efficient selective-ID secure identity-based encryption without random oracles. In Cachin and Camenisch [51], pages 223–238.

33. Dan Boneh and Xavier Boyen. Secure identity based encryption without random oracles. In Franklin [81], pages 443–459.

34. Dan Boneh and Xavier Boyen. Short signatures without random oracles. In Cachin and Camenisch [51], pages 56–73.

35. Dan Boneh, Xavier Boyen, and Eu-Jin Goh. Hierarchical identity based encryption with constant size ciphertext. In Cramer [73], pages 440–456.

36. Dan Boneh, Xavier Boyen, and Eu-Jin Goh. Hierarchical identity based encryption with constant size ciphertext. Cryptology ePrint Archive, Report 2005/015, 2005. http://eprint.iacr.org/.

37. Dan Boneh, Xavier Boyen, and Hovav Shacham. Short group signatures. In Franklin [81], pages 41–55.

38. Dan Boneh, Giovanni Di Crescenzo, Rafail Ostrovsky, and Giuseppe Persiano. Public key encryption with keyword search. In Cachin and Camenisch [51], pages 506–522.

39. Dan Boneh and Matthew K. Franklin. Identity-based encryption from the Weil pairing. *SIAM J. Comput.*, 32(3):586–615, 2003. Earlier version appeared in the proceedings of CRYPTO 2001.

40. Dan Boneh, Craig Gentry, and Michael Hamburg. Space-efficient identity based encryption without pairings. In *FOCS*, pages 647–657. IEEE Computer Society, 2007.

41. Dan Boneh, Craig Gentry, Ben Lynn, and Hovav Shacham. Aggregate and verifiably encrypted signatures from bilinear maps. In Biham [31], pages 416–432.

42. Dan Boneh, Craig Gentry, and Brent Waters. Collusion resistant broadcast encryption with short ciphertexts and private keys. In Shoup [161], pages 258–275.

43. Dan Boneh, Eu-Jin Goh, and Kobbi Nissim. Evaluating 2-DNF formulas on ciphertexts. In Joe Kilian, editor, *TCC*, volume 3378 of *Lecture Notes in Computer Science*, pages 325–341. Springer, 2005.

44. Dan Boneh and Jonathan Katz. Improved efficiency for CCA-secure cryptosystems built using identity-based encryption. In Alfred Menezes, editor, *CT-RSA*, volume 3376 of *Lecture Notes in Computer Science*, pages 87–103. Springer, 2005.

45. Dan Boneh, Ben Lynn, and Hovav Shacham. Short signatures from the Weil pairing. In Colin Boyd, editor, *ASIACRYPT*, volume 2248 of *Lecture Notes in Computer Science*, pages 514–532. Springer, 2001.

46. Xavier Boyen. General ad hoc encryption from exponent inversion IBE. In Moni Naor, editor, *EUROCRYPT*, volume 4515 of *Lecture Notes in Computer Science*, pages 394–411. Springer, 2007.

47. Xavier Boyen. The uber-assumption family. In Steven D. Galbraith and Kenneth G. Paterson, editors, *Pairing*, volume 5209 of *Lecture Notes in Computer Science*, pages 39–56. Springer, 2008.

48. Xavier Boyen and Luther Martin. Identity-based cryptography standard (IBCS) #1, December 2007. IETF memo.

49. Xavier Boyen, Qixiang Mei, and Brent Waters. Direct chosen ciphertext security from identity-based techniques. In Vijay Atluri, Catherine Meadows, and Ari Juels, editors, *ACM Conference on Computer and Communications Security*, pages 320–329. ACM, 2005.

50. Mike Burmester and Yvo Desmedt. A secure and scalable group key exchange system. *Inf. Process. Lett.*, 94(3):137–143, 2005.

51. Christian Cachin and Jan Camenisch, editors. *Advances in Cryptology - EUROCRYPT 2004, International Conference on the Theory and Applications of Cryptographic Techniques, Interlaken, Switzerland, May 2-6, 2004, Proceedings*, volume 3027 of *Lecture Notes in Computer Science*. Springer, 2004.

52. Ran Canetti, Oded Goldreich, and Shai Halevi. The random oracle methodology, revisited. *J. ACM*, 51(4):557–594, 2004.

53. Ran Canetti, Shai Halevi, and Jonathan Katz. A forward-secure public-key encryption scheme. In Biham [31], pages 255–271.

54. Ran Canetti, Shai Halevi, and Jonathan Katz. Chosen-ciphertext security from identity-based encryption. In Cachin and Camenisch [51], pages 207–222.
55. David Cash, Dennis Hofheinz, and Eike Kiltz. How to delegate a lattice basis. Cryptology ePrint Archive, Report 2009/351, 2009. http://eprint.iacr.org/.
56. David Cash, Dennis Hofheinz, Eike Kiltz, and Chris Peikert. Bonsai trees, or how to delegate a lattice basis. In Gilbert [95], pages 523–552.
57. Debrup Chakraborty and Palash Sarkar. A general construction of tweakable block ciphers and different modes of operations. *IEEE Transactions on Information Theory*, 54(5):1991–2006, 2008.
58. Sanjit Chatterjee, Darrel Hankerson, and Alfred Menezes. On the efficiency and security of pairing-based protocols in the Type 1 and Type 4 settings. In M. Anwar Hasan and Tor Helleseth, editors, *WAIFI*, volume 6087 of *Lecture Notes in Computer Science*, pages 114–134. Springer, 2010.
59. Sanjit Chatterjee and Alfred Menezes. On cryptographic protocols employing asymmetric pairings – the role of ψ revisited. Cryptology ePrint Archive, Report 2009/480, 2009. http://eprint.iacr.org/.
60. Sanjit Chatterjee and Palash Sarkar. Trading time for space: towards an efficient IBE scheme with short(er) public parameters in the standard model. In Dong Ho Won and Seungjoo Kim, editors, *ICISC*, volume 3935 of *Lecture Notes in Computer Science*, pages 424–440. Springer, 2005.
61. Sanjit Chatterjee and Palash Sarkar. HIBE with short public parameters without random oracle. In X. Lai and K. Chen, editors, *ASIACRYPT*, volume 4284 of *Lecture Notes in Computer Science*, pages 145–160. Springer, 2006. see also Cryptology ePrint Archive, Report 2006/279, http://eprint.iacr.org/.
62. Sanjit Chatterjee and Palash Sarkar. New constructions of constant size ciphertext hibe without random oracle. In M.S. Rhee and B. Lee, editors, *ICISC*, volume 4296 of *Lecture Notes in Computer Science*, pages 310–327. Springer, 2006.
63. Sanjit Chatterjee and Palash Sarkar. Practical hybrid (hierarchical) identity-based encryption schemes based on the decisional bilinear diffie-hellman assumption. CACR technical report, Report 2010/020, 2010. http://www.cacr.math.uwaterloo.ca/tech_reports.html.
64. Liqun Chen, Zhaohui Cheng, and Nigel P. Smart. Identity-based key agreement protocols from pairings. *Int. J. Inf. Sec.*, 6(4):213–241, 2007.
65. Sherman S. M. Chow. Token-controlled public key encryption in the standard model. In Juan A. Garay, Arjen K. Lenstra, Masahiro Mambo, and René Peralta, editors, *ISC*, volume 4779 of *Lecture Notes in Computer Science*, pages 315–332. Springer, 2007.
66. Sherman S. M. Chow. Certificateless encryption. In M. Joye and G. Neven, editors, *Identity-Based Cryptography*, pages 135–155. IOS Press, 2008.
67. Sherman S. M. Chow. Removing escrow from identity-based encryption. In Jarecki and Tsudik [110], pages 256–276.
68. Sherman S. M. Chow. *New Privacy-Preserving Architectures for Identity-/Attribute-Based Encryption*. PhD thesis, New York University, 2010.
69. Sherman S. M. Chow, Colin Boyd, and Juan Manuel González Nieto. Security-mediated certificateless cryptography. In Moti Yung, Yevgeniy Dodis, Aggelos Kiayias, and Tal Malkin, editors, *Public Key Cryptography*, volume 3958 of *Lecture Notes in Computer Science*, pages 508–524. Springer, 2006.
70. Clifford Cocks. An identity based encryption scheme based on quadratic residues. In Bahram Honary, editor, *IMA Int. Conf.*, volume 2260 of *Lecture Notes in Computer Science*, pages 360–363. Springer, 2001.
71. Jean-Sébastien Coron. On the exact security of full domain hash. In Mihir Bellare, editor, *CRYPTO*, volume 1880 of *Lecture Notes in Computer Science*, pages 229–235. Springer, 2000.
72. Jean-Sébastien Coron. A variant of Boneh-Franklin IBE with a tight reduction in the random oracle model. *Des. Codes Cryptography*, 50(1):115–133, 2009.

73. Ronald Cramer, editor. *Advances in Cryptology - EUROCRYPT 2005, 24th Annual International Conference on the Theory and Applications of Cryptographic Techniques, Aarhus, Denmark, May 22-26, 2005, Proceedings*, volume 3494 of *Lecture Notes in Computer Science*. Springer, 2005.

74. Ronald Cramer and Victor Shoup. Design and analysis of practical public-key encryption schemes secure against adaptive chosen ciphertext attack. *SIAM J. Comput.*, 33(1):167–226, 2003.

75. A. Menezes D. Hankerson and M. Scott. Software implementation of pairings. In M. Joye and G. Neven, editors, *Identity-Based Cryptography*, pages 188–206. IOS Press, 2008.

76. Alexander W. Dent. A survey of certificateless encryption schemes and security models. *Int. J. Inf. Sec.*, 7(5):349–377, 2008.

77. Whitfield Diffie and Martin E. Hellman. New directions in cryptography. *IEEE Tr. Inf. Th.*, 22:644–654, 1976.

78. Yevgeniy Dodis and Nelly Fazio. Public key broadcast encryption for stateless receivers. In Joan Feigenbaum, editor, *Digital Rights Management Workshop*, volume 2696 of *Lecture Notes in Computer Science*, pages 61–80. Springer, 2002.

79. Taher Elgamal. A public key cryptosystem and a signature scheme based on discrete logarithms. *IEEE Transactions on Information Theory*, 31(4):469–472, 1985.

80. Amos Fiat and Moni Naor. Broadcast encryption. In Douglas R. Stinson, editor, *CRYPTO*, volume 773 of *Lecture Notes in Computer Science*, pages 480–491. Springer, 1993.

81. Matthew K. Franklin, editor. *Advances in Cryptology - CRYPTO 2004, 24th Annual International CryptologyConference, Santa Barbara, California, USA, August 15-19, 2004, Proceedings*, volume 3152 of *Lecture Notes in Computer Science*. Springer, 2004.

82. David Mandell Freeman. Converting pairing-based cryptosystems from composite-order groups to prime-order groups. In Gilbert [95], pages 44–61.

83. David Mandell Freeman, Michael Scott, and Edlyn Teske. A taxonomy of pairing-friendly elliptic curves. *Journal of Cryptology*, 23:224–280, 2010.

84. Gerhard Frey, Michael Müller, and Hans-Georg Rück. The Tate pairing and the discrete logarithm applied to elliptic curve cryptosystems. *IEEE Transactions on Information Theory*, 45(5):1717–1719, 1999.

85. Atsushi Fujioka, Katarou Suzuki, and Berkant Ustaoglu. Secure leakage resilient and efficient id-akes that can share identities private and master keys. In *Pairing*, 2010.

86. Eiichiro Fujisaki and Tatsuaki Okamoto. Secure integration of asymmetric and symmetric encryption schemes. In Michael J. Wiener, editor, *CRYPTO*, volume 1666 of *Lecture Notes in Computer Science*, pages 537–554. Springer, 1999.

87. Stephen D. Galbraith. Pairings, Chapter IX. In I. Blake, G. Seroussi, and N. Smart, editors, *Advances in Elliptic Curve Cryptography 2*. Cambridge University Press, 2008.

88. Steven D. Galbraith, Kenneth G. Paterson, and Nigel P. Smart. Pairings for cryptographers. *Discrete Applied Mathematics*, 156(16):3113–3121, 2008.

89. David Galindo. Boneh-Franklin identity based encryption revisited. In Luís Caires, Giuseppe F. Italiano, Luís Monteiro, Catuscia Palamidessi, and Moti Yung, editors, *ICALP*, volume 3580 of *Lecture Notes in Computer Science*, pages 791–802. Springer, 2005.

90. Craig Gentry. Certificate-based encryption and the certificate revocation problem. In Biham [31], pages 272–293.

91. Craig Gentry. Practical identity-based encryption without random oracles. In Vaudenay [166], pages 445–464.

92. Craig Gentry and Shai Halevi. Hierarchical identity based encryption with polynomially many levels. In Omer Reingold, editor, *TCC*, volume 5444 of *Lecture Notes in Computer Science*, pages 437–456. Springer, 2009.

93. Craig Gentry, Chris Peikert, and Vinod Vaikuntanathan. Trapdoors for hard lattices and new cryptographic constructions. In Richard E. Ladner and Cynthia Dwork, editors, *STOC*, pages 197–206. ACM, 2008.

94. Craig Gentry and Alice Silverberg. Hierarchical ID-based cryptography. In Yuliang Zheng, editor, *ASIACRYPT*, volume 2501 of *Lecture Notes in Computer Science*, pages 548–566. Springer, 2002.

95. Henri Gilbert, editor. *Advances in Cryptology - EUROCRYPT 2010, 29th Annual International Conference on the Theory and Applications of Cryptographic Techniques, French Riviera, May 30 - June 3, 2010. Proceedings*, volume 6110 of *Lecture Notes in Computer Science*. Springer, 2010.
96. Virgil D. Gligor and Pompiliu Donescu. Fast encryption and authentication: XCBC encryption and XECB authentication modes. In Mitsuru Matsui, editor, *FSE*, volume 2355 of *Lecture Notes in Computer Science*, pages 92–108. Springer, 2001.
97. Oded Goldreich, Shafi Goldwasser, and Shai Halevi. Public-key cryptosystems from lattice reduction problems. In Kaliski-Jr. [116], pages 112–131.
98. Oded Goldreich, Yoad Lustig, and Moni Naor. On chosen ciphertext security of multiple encryptions. Cryptology ePrint Archive, Report 2002/089, 2002. http://eprint.iacr.org/.
99. Shafi Goldwasser and Silvio Micali. Probabilistic encryption. *J. Comput. Syst. Sci.*, 28(2):270–299, 1984.
100. Vipul Goyal. Reducing trust in the PKG in identity based cryptosystems. In Alfred Menezes, editor, *CRYPTO*, volume 4622 of *Lecture Notes in Computer Science*, pages 430–447. Springer, 2007.
101. Vipul Goyal, Steve Lu, Amit Sahai, and Brent Waters. Black-box accountable authority identity-based encryption. In Peng Ning, Paul F. Syverson, and Somesh Jha, editors, *ACM Conference on Computer and Communications Security*, pages 427–436. ACM, 2008.
102. Shai Halevi, editor. *Advances in Cryptology - CRYPTO 2009, 29th Annual International Cryptology Conference, Santa Barbara, CA, USA, August 16-20, 2009. Proceedings*, volume 5677 of *Lecture Notes in Computer Science*. Springer, 2009.
103. Darrel Hankerson, Alfred Menezes, and Scott Vanstone. *Guide to Elliptic Curve Cryptography*. Springer, 2004.
104. Florian Hess, Nigel P. Smart, and Frederik Vercauteren. The Eta pairing revisited. *IEEE Transactions on Information Theory*, 52(10):4595–4602, 2006.
105. Jeffrey Hoffstein, Jill Pipher, and Joseph H. Silverman. NTRU: A ring-based public key cryptosystem. In Joe Buhler, editor, *ANTS*, volume 1423 of *Lecture Notes in Computer Science*, pages 267–288. Springer, 1998.
106. Jeremy Horwitz and Ben Lynn. Toward hierarchical identity-based encryption. In Lars R. Knudsen, editor, *EUROCRYPT*, volume 2332 of *Lecture Notes in Computer Science*, pages 466–481. Springer, 2002.
107. Thomas Icart. How to hash into elliptic curves. In Halevi [102], pages 303–316.
108. IEEE-P1363.3. IEEE P1636.3TM/D1 draft standard for identity-based public-key cryptography using pairings, April 2008.
109. T. Itoh and S. Tsujii. A fast algorithm for computing multiplicative inverses in $GF(2^m)$ using normal bases. *Information and Computation*, 83, 1989.
110. Stanisław Jarecki and Gene Tsudik, editors. *Public Key Cryptography - PKC 2009, 12th International Conference on Practice and Theory in Public Key Cryptography, Irvine, CA, USA, March 18-20, 2009. Proceedings*, volume 5443 of *Lecture Notes in Computer Science*. Springer, 2009.
111. Mahabir Prasad Jhanwar and Rana Barua. A variant of Boneh-Gentry-Hamburg's pairing-free identity based encryption scheme. In Moti Yung, Peng Liu, and Dongdai Lin, editors, *Inscrypt*, volume 5487 of *Lecture Notes in Computer Science*, pages 314–331. Springer, 2008.
112. Antoine Joux. A one round protocol for tripartite Diffie-Hellman. In Wieb Bosma, editor, *ANTS*, volume 1838 of *Lecture Notes in Computer Science*, pages 385–394. Springer, 2000.
113. Antoine Joux. A one round protocol for tripartite Diffie-Hellman. *J. Cryptology*, 17(4):263–276, 2004. Earlier version appeared in the proceedings of ANTS-IV.
114. Charanjit S. Jutla. Encryption modes with almost free message integrity. In Birgit Pfitzmann, editor, *EUROCRYPT*, volume 2045 of *Lecture Notes in Computer Science*, pages 529–544. Springer, 2001.
115. David Kahn. *The Codebreakers: The Comprehensive History of Secret Communication from Ancient Times to the Internet*. Scribner.

116. Burton S. Kaliski-Jr., editor. *Advances in Cryptology - CRYPTO '97, 17th Annual International Cryptology Conference, Santa Barbara, California, USA, August 17-21, 1997, Proceedings*, volume 1294 of *Lecture Notes in Computer Science*. Springer, 1997.
117. Aniket Kate and Ian Goldberg. Distributed private-key generators for identity-based cryptography. In Juan A. Garay and Roberto De Prisco, editors, *SCN*, volume 6280 of *Lecture Notes in Computer Science*, pages 436–453. Springer, 2010.
118. Jonathan Katz and Nan Wang. Efficiency improvements for signature schemes with tight security reductions. In Sushil Jajodia, Vijayalakshmi Atluri, and Trent Jaeger, editors, *ACM Conference on Computer and Communications Security*, pages 155–164. ACM, 2003.
119. Jonathan Katz and Moti Yung. Complete characterization of security notions for probabilistic private-key encryption. In *STOC*, pages 245–254, 2000.
120. Eike Kiltz and David Galindo. Direct chosen-ciphertext secure identity-based key encapsulation without random oracles. In Batten and Safavi-Naini [20], pages 336–347. full version available at http://eprint.iacr.org/2006/034.
121. Eike Kiltz and Yevgeniy Vahlis. CCA2 secure IBE: Standard model efficiency through authenticated symmetric encryption. In Tal Malkin, editor, *CT-RSA*, volume 4964 of *Lecture Notes in Computer Science*, pages 221–238. Springer, 2008.
122. Neal Koblitz. Elliptic curve cryptosystems. *Mathematics of Computation*, 48:203–209, 1987.
123. Neal Koblitz. Constructing elliptic curve cryptosystems in characteristic 2. In Alfred Menezes and Scott A. Vanstone, editors, *CRYPTO*, volume 537 of *Lecture Notes in Computer Science*, pages 156–167. Springer, 1990.
124. Neal Koblitz and Alfred Menezes. Another look at "Provable Security". *J. Cryptology*, 20(1):3–37, 2007.
125. Hugo Krawczyk, editor. *Advances in Cryptology - CRYPTO '98, 18th Annual International Cryptology Conference, Santa Barbara, California, USA, August 23-27, 1998, Proceedings*, volume 1462 of *Lecture Notes in Computer Science*. Springer, 1998.
126. Thomas Kuhn. *Structure of Scientific Revolutions*. Chicago University Press, 1962.
127. Kaoru Kurosawa and Yvo Desmedt. A new paradigm of hybrid encryption scheme. In Franklin [81], pages 426–442.
128. E. Lee, H.-S. Lee, and C.-M. Park. Efficient and generalized pairing computation on abelian varieties. *IEEE Transactions on Information Theory*, 55(4):1793–1803, 2009.
129. A. K. Lenstra and H. W. Lenstra, editors. *The development of the number field sieve*. Springer-Verlag, 1993.
130. Arjen K. Lenstra and Eric R. Verheul. Selecting cryptographic key sizes. *J. Cryptology*, 14(4):255–293, 2001.
131. Allison B. Lewko and Brent Waters. New techniques for dual system encryption and fully secure HIBE with short ciphertexts. In Daniele Micciancio, editor, *TCC*, volume 5978 of *Lecture Notes in Computer Science*, pages 455–479. Springer, 2010.
132. R. Lidl and H. Niederreiter. *Introduction to finite fields and their applications, revised edition*. Cambridge University Press, 1994.
133. Alfred Menezes, Tatsuaki Okamoto, and Scott A. Vanstone. Reducing elliptic curve logarithms to logarithms in a finite field. *IEEE Transactions on Information Theory*, 39(5):1639–1646, 1993.
134. Victor S. Miller. Use of elliptic curves in cryptography. In Hugh C. Williams, editor, *CRYPTO*, volume 218 of *Lecture Notes in Computer Science*, pages 417–426. Springer, 1985.
135. Victor S. Miller. Short programs for functions on curves. Unpublished manuscript, 1986. http://crypto.stanford.edu/miller/miller.pdf.
136. Victor S. Miller. The Weil pairing, and its efficient calculation. *J. Cryptology*, 17(4):235–261, 2004.
137. David Naccache. Secure and practical identity-based encryption. *IET Information Security*, 1(2):59–64, 2007.
138. Michael Naehrig, Ruben Niederhagen, and Peter Schwabe. New software speed records for cryptographic pairings. In Michel Abdalla and Paulo S. L. M. Barreto, editors, *LATINCRYPT*, volume 6212 of *Lecture Notes in Computer Science*, pages 109–123. Springer, 2010.

139. Dalit Naor, Moni Naor, and Jeffery Lotspiech. Revocation and tracing schemes for state-less receivers. In Joe Kilian, editor, *CRYPTO*, volume 2139 of *Lecture Notes in Computer Science*, pages 41–62. Springer, 2001.

140. Moni Naor and Moti Yung. Universal one-way hash functions and their cryptographic applications. In *STOC*, pages 33–43. ACM, 1989.

141. V. I. Nechaev. Complexity of a determinate algorithm for the discrete logarithm. *Mathematical Notes*, 55(2):165–172, 1994.

142. Phong Nguyen and Brigitte Vallée, editors. *The LLL algorithm: survey and applications*. Springer, 2009.

143. Leonardo B. Oliveira, Diego F. Aranha, Eduardo Morais, Felipe Daguano, Julio López, and Ricardo Dahab. TinyTate: Computing the Tate pairing in resource-constrained sensor nodes. In *NCA*, pages 318–323. IEEE Computer Society, 2007.

144. Kenneth G. Paterson and Jacob C. N. Schuldt. Efficient identity-based signatures secure in the standard model. In Batten and Safavi-Naini [20], pages 207–222.

145. Kenneth G. Paterson and Sriramkrishnan Srinivasan. On the relations between non-interactive key distribution, identity-based encryption and trapdoor discrete log groups. *Des. Codes Cryptography*, 52(2):219–241, 2009.

146. Chris Peikert. Bonsai trees (or, arboriculture in lattice-based cryptography). Cryptology ePrint Archive, Report 2009/359, 2009. http://eprint.iacr.org/.

147. Oded Regev. On lattices, learning with errors, random linear codes, and cryptography. In Harold N. Gabow and Ronald Fagin, editors, *STOC*, pages 84–93. ACM, 2005.

148. Ronald L. Rivest, Adi Shamir, and Leonard M. Adleman. A method for obtaining digital signatures and public-key cryptosystems. *Commun. ACM*, 21(2):120–126, 1978.

149. Phillip Rogaway. Efficient instantiations of tweakable blockciphers and refinements to modes OCB and PMAC. In Pil Joong Lee, editor, *ASIACRYPT*, volume 3329 of *Lecture Notes in Computer Science*, pages 16–31. Springer, 2004.

150. Amit Sahai and Brent Waters. Fuzzy identity-based encryption. In Cramer [73], pages 457–473.

151. Ryuichi Sakai, Kiyoshi Ohgishi, and Masao Kasahara. Cryptosystems based on pairing. In *Symposium on Cryptography and Information Security – SCIS*, 2000. In Japanese, English version available from the authors.

152. Palash Sarkar. Masking-based domain extenders for UOWHFs: bounds and constructions. *IEEE Transactions on Information Theory*, 51(12):4299–4311, 2005.

153. Palash Sarkar. Pseudo-random functions and parallelizable modes of operations of a block cipher. *IEEE Transactions on Information Theory*, 2010. to appear.

154. Palash Sarkar and Sanjit Chatterjee. Construction of a hybrid HIBE protocol secure against adaptive attacks. In Willy Susilo, Joseph K. Liu, and Yi Mu, editors, *ProvSec*, volume 4784 of *Lecture Notes in Computer Science*, pages 51–67. Springer, 2007.

155. Adi Shamir. Identity-based cryptosystems and signature schemes. In G. R. Blakley and David Chaum, editors, *CRYPTO*, volume 196 of *Lecture Notes in Computer Science*, pages 47–53. Springer, 1984.

156. Adi Shamir. A polynomial-time algorithm for breaking the basic Merkle-Hellman cryptosystem. *IEEE Transactions on Information Theory*, 30(5):699–704, 1984.

157. Elaine Shi and Brent Waters. Delegating capabilities in predicate encryption systems. In Luca Aceto, Ivan Damgård, Leslie Ann Goldberg, Magnús M. Halldórsson, Anna Ingólfsdóttir, and Igor Walukiewicz, editors, *ICALP (2)*, volume 5126 of *Lecture Notes in Computer Science*, pages 560–578. Springer, 2008.

158. Victor Shoup. Lower bounds for discrete logarithms and related problems. In *EUROCRYPT*, pages 256–266, 1997.

159. Victor Shoup. A composition theorem for universal one-way hash functions. In Bart Preneel, editor, *EUROCRYPT*, volume 1807 of *Lecture Notes in Computer Science*, pages 445–452. Springer, 2000.

160. Victor Shoup. Sequences of games: a tool for taming complexity in security proofs. Cryptology ePrint Archive, Report 2004/332, 2004. http://eprint.iacr.org/.

161. Victor Shoup, editor. *Advances in Cryptology - CRYPTO 2005: 25th Annual International Cryptology Conference, Santa Barbara, California, USA, August 14-18, 2005, Proceedings*, volume 3621 of *Lecture Notes in Computer Science*. Springer, 2005.

162. Victor Shoup. A proposal for an ISO standard for public key encryption (version 2.1), December 20, 2001. available from http://www.shoup.net/papers/.

163. Joseph H. Silverman. *The Arithmetic of Elliptic Curves*. Springer, 1986.

164. Simon Singh. *The Code Book: The Science of Secrecy from Ancient Egypt to Quantum Cryptography*. Anchor.

165. Nigel P. Smart. Access control using pairing based cryptography. In Marc Joye, editor, *CT-RSA*, volume 2612 of *Lecture Notes in Computer Science*, pages 111–121. Springer, 2003.

166. Serge Vaudenay, editor. *Advances in Cryptology - EUROCRYPT 2006, 25th Annual International Conference on the Theory and Applications of Cryptographic Techniques, St. Petersburg, Russia, May 28 - June 1, 2006, Proceedings*, volume 4004 of *Lecture Notes in Computer Science*. Springer, 2006.

167. F. Vercauteren. Optimal pairings. Cryptology ePrint Archive, Report 2008/096, 2008. http://eprint.iacr.org/.

168. Yodai Watanabe, Junji Shikata, and Hideki Imai. Equivalence between semantic security and indistinguishability against chosen ciphertext attacks. In Yvo Desmedt, editor, *Public Key Cryptography*, volume 2567 of *Lecture Notes in Computer Science*, pages 71–84. Springer, 2003.

169. Brent Waters. Efficient identity-based encryption without random oracles. In Cramer [73], pages 114–127.

170. Brent Waters. Dual system encryption: Realizing fully secure IBE and HIBE under simple assumptions. In Halevi [102], pages 619–636.

171. Brent Waters. Dual system encryption: Realizing fully secure ibe and hibe under simple assumptions. Cryptology ePrint Archive, Report 2009/385, 2009.

172. Danfeng Yao, Nelly Fazio, Yevgeniy Dodis, and Anna Lysyanskaya. ID-based encryption for complex hierarchies with applications to forward security and broadcast encryption. In Vijayalakshmi Atluri, Birgit Pfitzmann, and Patrick Drew McDaniel, editors, *ACM Conference on Computer and Communications Security*, pages 354–363. ACM, 2004.

Index